P9-AZX-197

earth sheltered housing design

guidelines, examples, and references

prepared by: the underground space center
university of minnesota

prepared for: the minnesota energy agency

funded by: the legislative commission on minnesota resources

The financial support of the Minnesota Energy Agency is acknowledged, but the authors assume complete responsibility for the contents herein.

VNR VAN NOSTRAND REINHOLD COMPANY
New York Cincinnati Toronto London Melbourne

Library of Congress Catalog Card Number 78-25555
ISBN 0-442-28821-2 paperback
ISBN 0-442-26157-8 cloth

Printed in the United States of America

Published in 1979 by Van Nostrand Reinhold Company
A division of Litton Educational Publishing, Inc.
135 West 50th Street, New York, NY 10020, U.S.A.

Van Nostrand Reinhold Limited
1410 Birchmount Road
Scarborough, Ontario M1P 2E7, Canada

Van Nostrand Reinhold Australia Pty. Ltd.
17 Queen Street
Mitcham, Victoria 3132, Australia

Van Nostrand Reinhold Company Limited
Molly Millars Lane
Wokingham, Berkshire, England

16 15 14 13 12 11 10 9 8 7 6 5 4 3

Library of Congress Cataloging in Publication Data
Minnesota. University. Underground Space Center.
Earth sheltered housing design.

Bibliography: p.
1. Earth sheltered houses. I. Minnesota. Energy Agency. II. Title.
TH4819.E27M56 1978 728 78-25555
ISBN 0-442-28821-2 p
ISBN 0-442-26157-8 c

Legislative Commission on Minnesota Resources

B-46 STATE CAPITOL ST. PAUL, MINNESOTA 55155 (612) 296-2406

ROBERT E. HANSEN
EXECUTIVE DIRECTOR

To the Citizens of Minnesota

The Legislative Commission on Minnesota Resources welcomes the appearance of this publication. The concept of earth sheltered housing is rapidly developing as a viable alternative for Minnesota to reduce the dependence of housing on an uninterrupted supply of fossil fuel energy. This problem is of great concern to the Commission, which has been pleased to fund the earth sheltered housing design study as part of a series of grants for research in energy conservation made through the Minnesota Energy Agency.

The Commission hopes the information generated by this study will help the large numbers of people who have expressed interest in this concept convert their ideas into a workable reality. The state desperately needs more energy efficient housing so that we can conserve our dwindling conventional energy resources while we attempt to develop some alternatives.

Sincerely,

Rep. James R. Casserly,
Chairman, LCMR

preface

This study has arisen out of a long standing interest at the University of Minnesota in the greater use of underground space. It is perhaps ironic that the initial tendency of our involvement was to play down the concept of underground housing and to concentrate on underground construction as it applied to larger buildings and facilities. The ideas proposed were primarily aimed at saving our precious land surface for only those activities for which the surface is essential or the most desirable, i.e., people, agriculture and parks on the surface—warehouses and manufacturing plants underground.

It was thought at the time that mentioning the possibility of underground or earth sheltered homes (because of the anticipated psychological connotations) would have a negative impact on the other ideas being proposed. In a sense this has been true, but not at all in the manner expected. Thanks, in great part, to Thomas Bligh's concepts and vigorous activity in the area of earth sheltered housing, the response to the idea has been such as to simply crowd out many of our other planned activities. This should not really have surprised us; the same advantages of land saving and environmental preservation also apply to housing and in addition, the smaller scale of the building allows a greater impact of the earth's temperature moderation on the structure's energy performance.

In fact, man and other living creatures have always turned to the earth for protection from the elements and extremes of climate. It is only in the historically brief era of plentiful and cheap fossil fuel supplies, that we have been able to design a house without regard to the climate and then to supply whatever equipment and energy may be necessary to keep us comfortable.

Now that fossil fuel supplies are dwindling and fuel prices rapidly rising, it appears time to reconsider what the earth has to offer. With standard modern construction techniques, there is no need for a return to cave dwelling. The goal of earth sheltered houses is to keep or improve the relationship to the outdoors and the comfort of conventional houses while pulling the earth as a blanket around as much of the house as possible. The earth then acts as a barrier to wind chill and unwanted infiltration as well as direct heat loss.

Few of the houses shown in this study could be classed as entirely underground but all use the earth, either to blend the house with its surroundings or to improve its energy performance—hence the term earth sheltering. Not only does earth sheltering give a great reduction in overall energy consumption under normal conditions but its most striking improvement is in the independence of the house from an uninterrupted source of fossil fuel energy, especially in survival conditions such as are experienced in Minnesota and other northern states in the middle of winter.

Ray Sterling
Director, The Underground Space Center

major contributors:

DR. RAY STERLING P.E. Underground Space Center Univ. of Minn. (Formerly with Setter, Leach, and Lindstrom, Mpls.)	Project Coordinator Structural Design Waterproofing Public Policy
JOHN CARMODY TOM ELLISON Carmody & Ellison Arch. Design 1800 Englewood Av. St. Paul, Minn.	Assoc. Coordinators Site Planning Arch. Design Graphics, Layout & Illustrations
PAUL SHIPP Dept. of Mech. Eng. Univ. of Minn.	Energy Use
TERRILL L. TILLMAN P.E. 208 Lawn Terrace Golden Valley, Minn. 55416	Energy Use
MARTIN LUNDE P.E. Bressler, Armitage & Lunde 1002 Wesley Temple Bldg. Mpls., Minn. 55403	Energy Use
DR. CHARLES NELSON P.E. Dept. of Civil & Min. Eng. Univ. of Minn.	Structural Design
KEN LABS Star Route Mechanicsville, PA. 18934	Gathered Information for Part B

principal investigators:

DR. CHARLES FAIRHURST
DR. THOMAS BLIGH
Dept of Civil & Min. Eng.
Univ. of Minn.

advisors:

RICHARD VASATKA P.E.
WILLIAM SCOTT A.I.A.
ARNOLD CISEWSKI P.E.
Setter, Leach & Lindstrom
1011 Nicollet Mall
Mpls., Minn. 55403

DAVID BENNETT A.I.A.
Myers & Bennett Studio, B.R.W.
7101 York Ave. So.
Edina, Minn. 55435

additional assistance:

BUD GIESEN P.E. — Structural Design
Setter, Leach & Lindstrom

JAMES FENLESON — Cost Estimating
Setter, Leach & Lindstrom

WESTLUND CONSTRUCTION CO. — Cost Estimating
1781 Hamline Ave.
St. Paul, Minn.

BRIAN McGROARTY — Waterproofing
J. D. Godward Co.
11000 Lyndale Ave. So.
Mpls., Minn. 55420

TERRY NICHOLS — Waterproofing
Nichols & Hines
94 Crocus Place
St. Paul, Minn.

WILLIAM KWASNY P.E. — Waterproofing
Soil Testing Services
2405 Annapolis Lane
Mpls., Minn.

ROBERT E. PENDERGAST P.E. — Site Investigation
Geotechnical Eng. Corp.
1925 Oakcrest Ave.
Roseville, Minn.

PETER HERZOG — Architecture Energy Use
Architect
Assoc. Energy Consultants
7505 W. Hwy. 7
Mpls., Minn. 55426

part a: design considerations

part b: existing earth-sheltered houses

part c: additional information

foreword

The intent of this study is to present information which will be useful in the architectural design of earth sheltered houses. Part A discusses design guidelines and includes pertinent factors to be considered. Part B gives plans, details and photographs of existing examples of earth sheltered houses from around the country. These serve to show a number of different ways in which the design constraints discussed in Part A have been dealt with in individual designs. Part C is intended to ease access to further detailed information and includes an annotated bibliography.

In part A, planning, architectural design and construction aspects are all written with the intent of giving the important considerations in house layout and overall design. The information in these sections will be somewhat elementary to professionals in these fields but it is intended to give an interested individual a grasp of the essential parameters affecting earth sheltered design. The provision of this design information should not be construed to mean that no outside assistance with design is necessary. In particular, the structural design for earth sheltered houses should not be treated lightly and professional assistance in this aspect should normally be sought.

It is hoped, on the other hand, that the information given on the energy performance of earth sheltered houses and the public policy issues relating to their construction will be of value to most readers because so little data has existed on these topics. The energy discussions, of necessity, become more technical than the other sections but every attempt has been made to allow an average reader to follow the general discussions of energy performance and mechanical equipment selection. Units in the sections dealing with energy use are generally given in S.I. units with the Imperial equivalents in brackets. This is in keeping with the general trend to metrication and also many of the units used are already familiar (°C, kilowatt-hour, for example). In other sections of the report where S.I. units have little practical meaning yet (such as in lumber sizes), Imperial units only are used.

Only single family dwellings are considered in this study although clearly most of the considerations would also be applicable for use in earth sheltered housing developments. Consideration of any special design issues relating to multiple housing developments will be an important next step in the development of design information. The study has concentrated on the needs and requirements for housing in Minnesota but much of the information presented will be of general application. The discussion of earth sheltered housing involves the integration of many other energy saving concepts such as passive solar design. The merging of these essentially passive systems can be part of a trend towards durable, low maintenance structures which will use as many natural elements as possible to provide a low total life cycle cost and energy performance.

part a:
design considerations

1 site considerations

- orientation
  - sun
  - wind
  - views
- topography
- vegetation
- lot size and adjacent structures
- soil and groundwater

introduction

Selection and planning of a site for an earth sheltered house is usually the first and often one of the most important aspects of the entire design process. For conventional above grade housing, site selection and planning are often routine matters since typical designs for such houses are well known and important energy saving considerations are often ignored. With any well designed energy efficient house it is important to clearly understand the impact of basic site considerations such as proper orientation as well as the location of vegetation on the site. In order to select and plan a site for an earth sheltered house, it is also important to understand various site considerations which are particularly relevant for this unique type of housing. These include the topography, soil and groundwater conditions, and the lot size and location of adjacent structures. Some of this information pertaining to sites is quite simple and obvious while some is rather complex and technical. This section of the study attempts to identify the basic site considerations for an earth sheltered home and briefly explains the impact of each item.

orientation

One major facet of site planning is the actual location and orientation of the structure. Although a house is referred to as earth sheltered, it should not be thought of as completely covered by earth since window and door openings are required and desirable for a number of reasons. The grouping of these openings and the direction which they face can be referred to as the orientation of the structure on the site. The three major determinants of orientation are sun, wind and outside views. Proper orientation of a house with respect to sun and wind can produce significant energy savings in addition to those inherent to earth sheltering. Exterior views are an important aesthetic and psychological determinant of orientation.

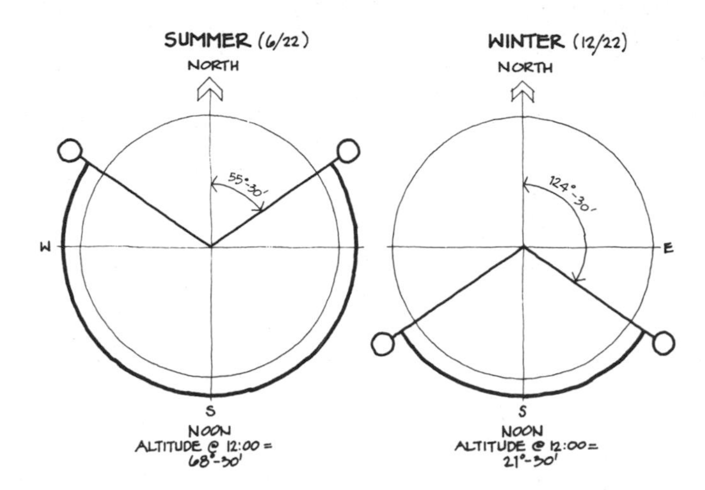

1-1 solar angles at 45° north latitude (mpls)

sun

The sun is one of the most important determinants in energy efficient building design. The radiant energy from the sun can be used in both an active and a passive manner to provide heat for a structure. Most types of active solar collector systems employ a flat plate collector which may be directly attached or located adjacent to the building. The use of an active solar collector system attached to the structure will have direct impact on the orientation and design of an earth sheltered residence. Usually the collector will be facing directly south, but this may vary depending on the particular system.

All passive solar collection methods are based on trapping the radiant energy of the sun which enters the house through the windows. The use of passive solar collection techniques in an energy efficient residence is a very desirable concept since it does not involve the capital expense that an active solar collector does and can provide a substantial amount of energy. According to a recent study on passive solar energy a double glazed window facing south will produce a net energy gain even without the use of drapes or shutters at night. This available radiant heat from the sun has a great impact on site orientation of earth sheltered design where the window openings are likely to be concentrated on one side of the house in order to maximize the earth cover. Considering sunlight alone, the best site orientation for any earth sheltered house would place all of the window openings on the south side with the remaining three sides completely earth covered. Passive solar collection is diminished considerably with east and west

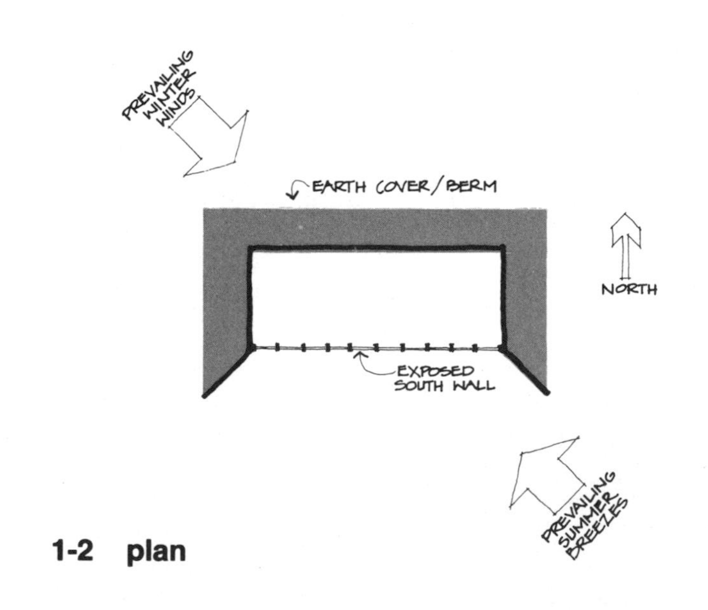

1-2 plan

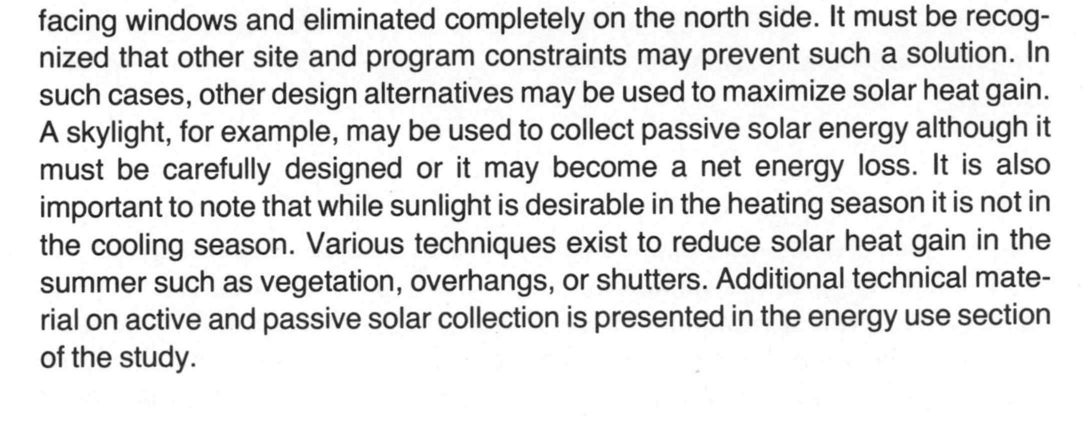

facing windows and eliminated completely on the north side. It must be recognized that other site and program constraints may prevent such a solution. In such cases, other design alternatives may be used to maximize solar heat gain. A skylight, for example, may be used to collect passive solar energy although it must be carefully designed or it may become a net energy loss. It is also important to note that while sunlight is desirable in the heating season it is not in the cooling season. Various techniques exist to reduce solar heat gain in the summer such as vegetation, overhangs, or shutters. Additional technical material on active and passive solar collection is presented in the energy use section of the study.

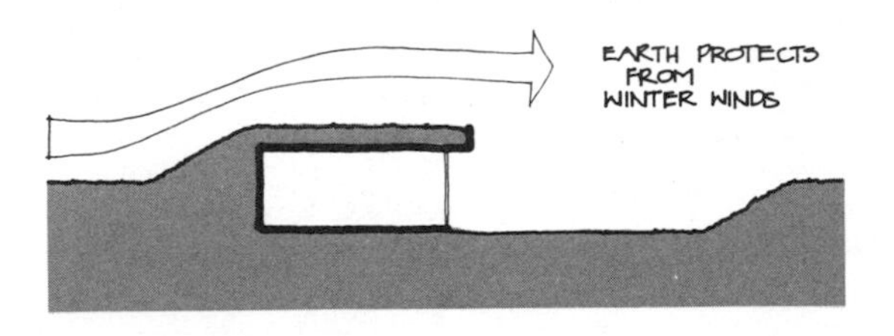

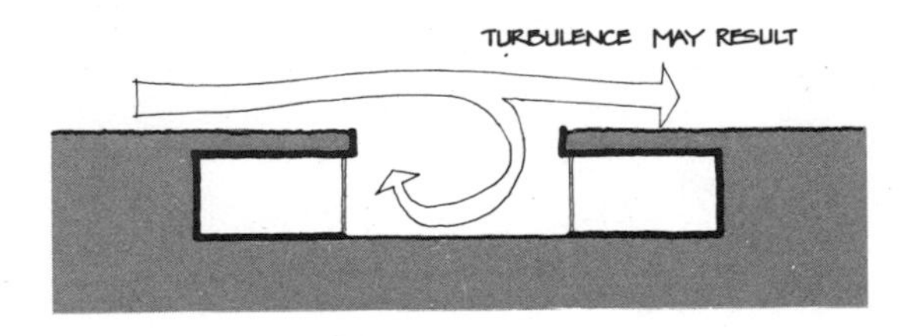

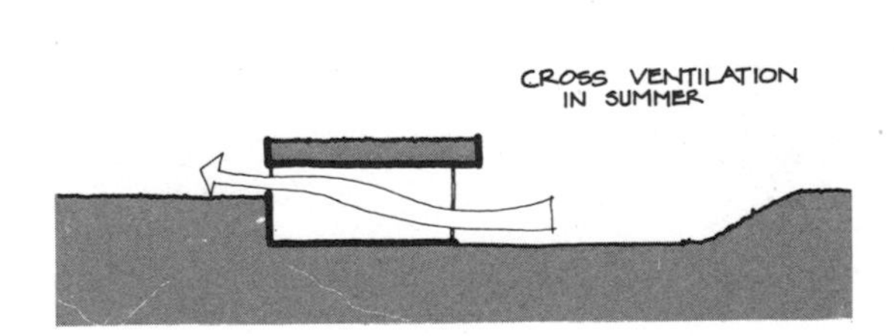

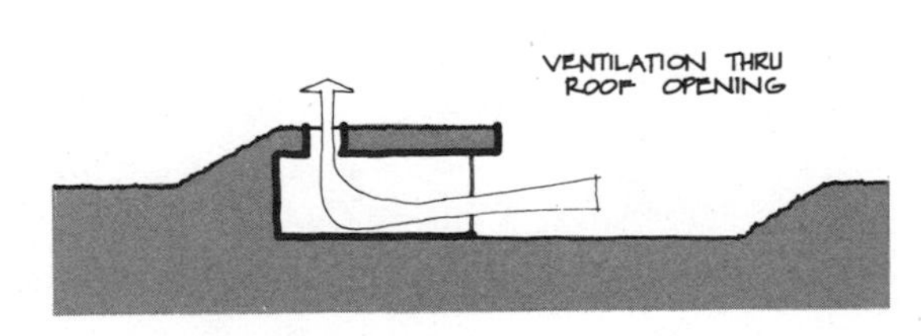

1-3 effects of wind

wind

The effect of wind on the orientation of an earth sheltered structure is a serious energy consideration. Since direct exposure to cold winter winds increases heat loss due to infiltration and a wind chill effect, it is desirable to protect a building as much as possible from this exposure. In the northern hemisphere the prevailing winter winds are from the northwest. Minimizing window and door openings on the north and west side of the house will enhance its energy performance. Earth sheltered construction offers a very unique opportunity to totally shield the structure from prevailing winter winds and use the earth to completely divert the wind over the structure. An earth sheltered design which includes a central courtyard may be substantially protected from prevailing winds, but minor wind turbulence may result. The behavior of this wind turbulence depends on many specific details of a design such as the size of the courtyards, the edge details, and the use of landscaping.

In the summer it is desirable to take advantage of prevailing breezes to provide natural ventilation. Generally the prevailing summer breezes are from the southeast, although this can vary from site to site depending on local conditions, trees, and topography. Unfortunately, the orientation of an earth sheltered structure with all window openings to the south will not create a well ventilated house. Some outlets such as windows or vents must be provided on the north side or the top of the house to create natural cross ventilation. There are many possible design variations which will result in good natural ventilation. If a designer wishes to maximize the earth cover at the expense of natural ventilation some form of mechanical ventilation can be provided. Issues relative to ventilation and human comfort are discussed in greater detail in the energy use section.

views

The presence of a desirable or undesirable view on a residential site is typically a major determinant of the window orientation. In designing an earth sheltered home on a site with a predominant scenic view, there are two important considerations. First, if the desirable view is not to the south, then the orientation of major window openings toward the view will result in less potential heat gain from the sun. The second consideration is the ability to clearly see an exterior view from an earth sheltered house which may be lower than a typical house. This may not be a problem on a sloping site, but on a flat site there are several design alternatives which allow a clear exterior view from an earth sheltered space. These are discussed in the architectural design section.

In the case of an undesirable view such as a highway or an adjacent building, the orientation of major window openings away from the view is important. Earth sheltered designs may be particularly effective in screening out undesirable views by orienting windows into courtyards which are virtually isolated from the surrounding environment.

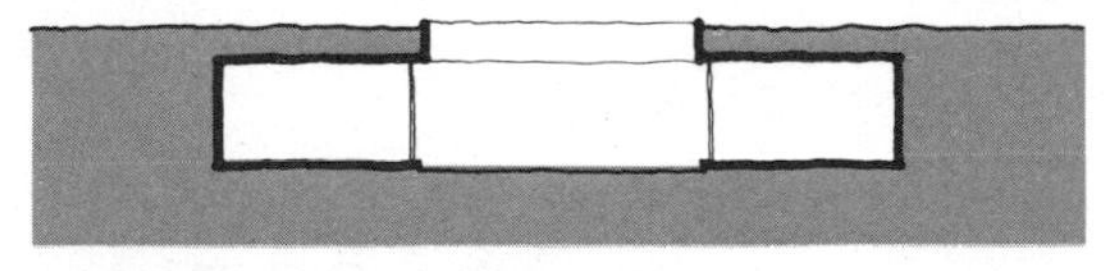

FLAT SITE – FULLY RECESSED

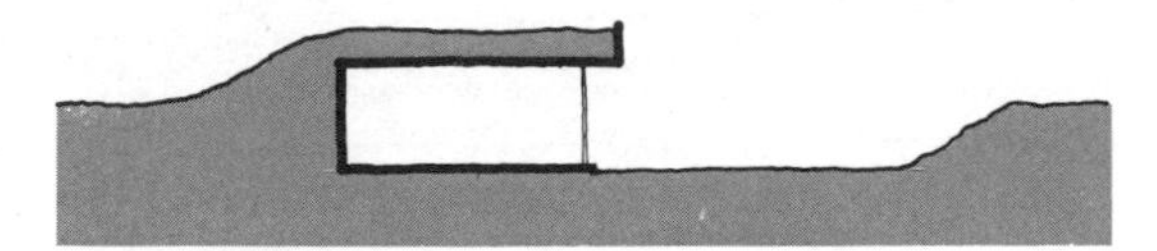

FLAT SITE – SEMI-RECESSED

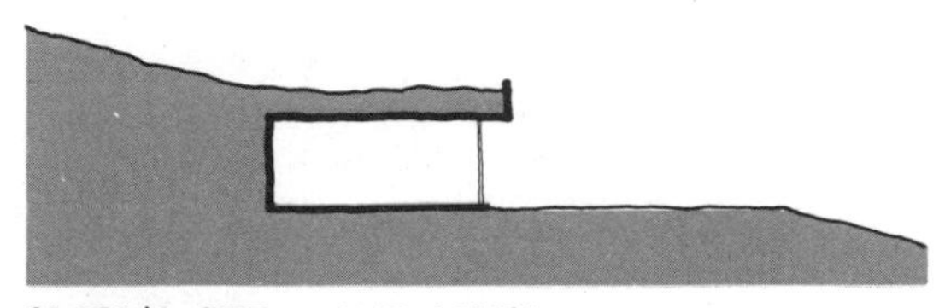

SLOPING SITE – ONE LEVEL

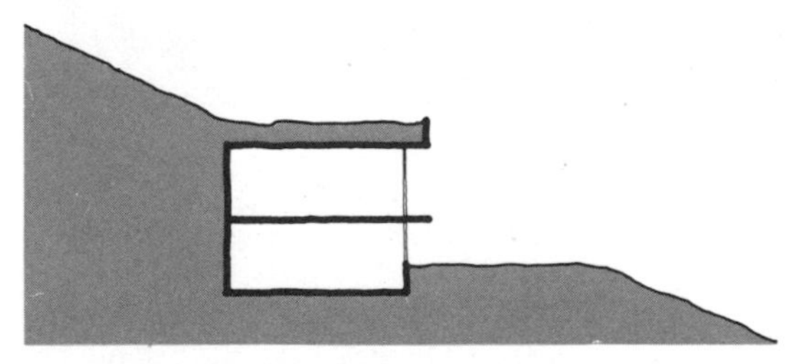

SLOPING SITE – TWO LEVELS

1-5 effects of topography

The topography of a building site can affect the design in a number of ways. Changes in terrain can directly affect the wind patterns and temperature around a building and certainly has a great impact on patterns of water runoff. However, the single most important effect of topography on an earth sheltered design is simply whether a site is flat or sloped and the degree and orientation of any slope. Since window openings are required by building codes and desirable for light and view, it can be assumed that earth sheltered houses are likely to be at least partially exposed to the outdoors.

On a predominantly flat site, a fully or semi-recessed design is possible, usually limited to one level below grade. A flat site may be used for a two level earth covered design, but a great deal of fill may be required depending, of course, on the specific design. A sloping site offers the opportunity to set an earth covered space into the hillside, however, the orientation of the window wall is then determined by the direction of the hillside. It is generally more desirable to work with a site sloping downward to the south, but it may be necessary to compromise maximum solar exposure on some sites. Also, a relatively steeply sloping site is more easily adaptable to a design with two levels of earth covered space. There are some significiant energy savings with a two level design which are discussed in the energy use and illustrative design sections of the study.

vegetation

Trees and shrubs on a building site have uses too numerous to mention ranging from symbolic and aesthetic uses to erosion and noise control. However, in the context of this study the existence of trees and shrubs on a site as well as the planting of new vegetation can be considered as another potential energy saving feature. Although exact amounts of energy savings are difficult to predict, there are uses of vegetation that are known to have a significant impact on energy performance. One effective use of trees in an earth sheltered design is to shade windows on the south side of the house in the summer when direct solar radiation is undesirable. If deciduous trees are used on the south side, they drop their leaves and allow the solar radiation to reach the windows in the winter when it is most needed. A second important use of vegetation to reduce energy consumption is wind control. In an earth sheltered design the earth berms may provide all of the necessary wind protection on the north and west sides. However, if any window openings are exposed to winter winds, evergreen plant materials can contribute to significant energy savings. In a report prepared for the Minnesota Energy Agency, a study in South Dakota is referred to which compares two identical houses, one with plants on three sides of the house and one with no windbreaks. The house with the windbreaks experienced a fuel reduction of 40%. This not only points out the effectiveness of trees and shrubs in reducing energy costs but it also indicates the potential savings that could be projected for a house that was earth sheltered on three sides providing a far more complete windbreak.

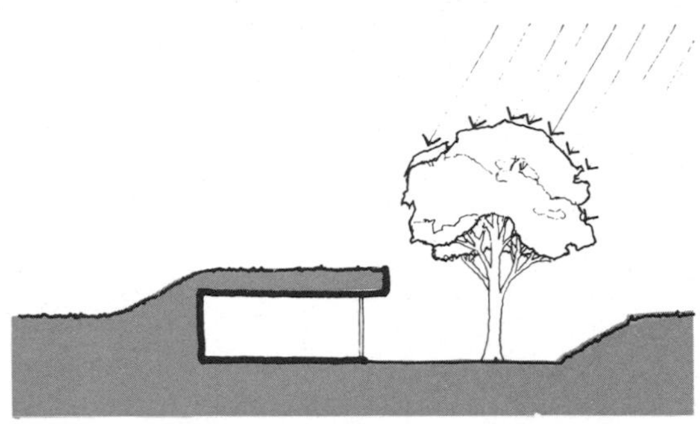

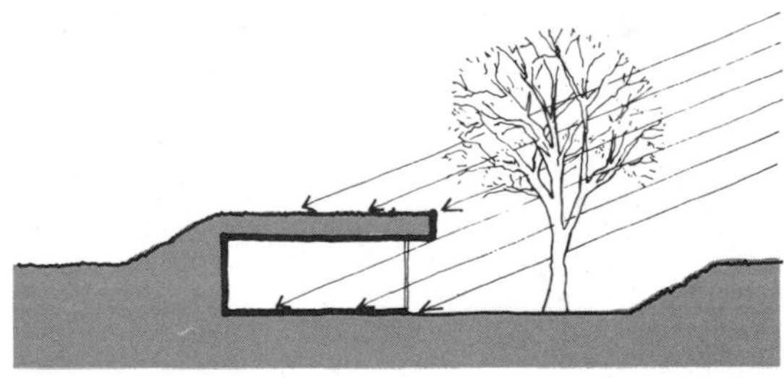

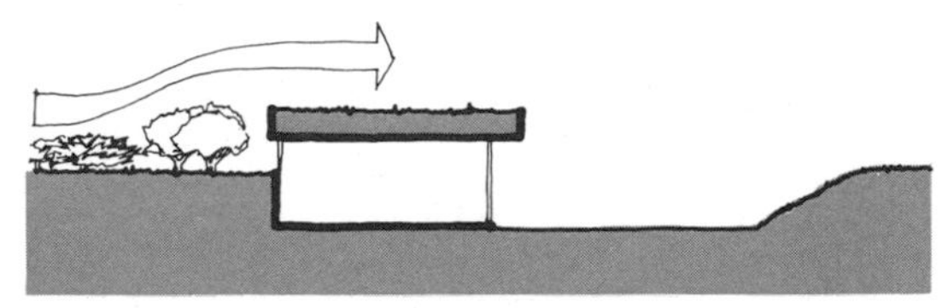

1-6 effects of vegetation

lot size and adjacent structures

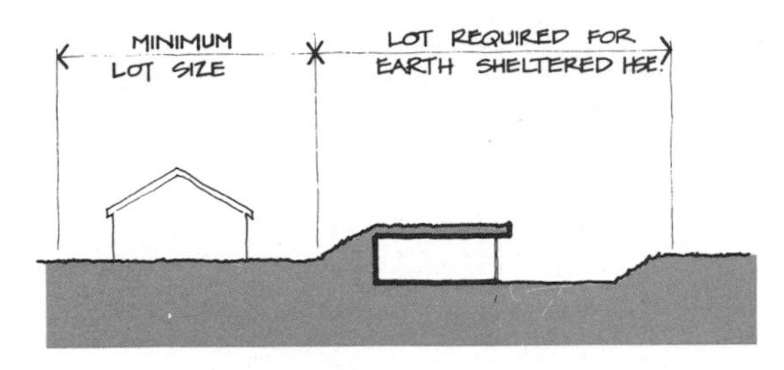

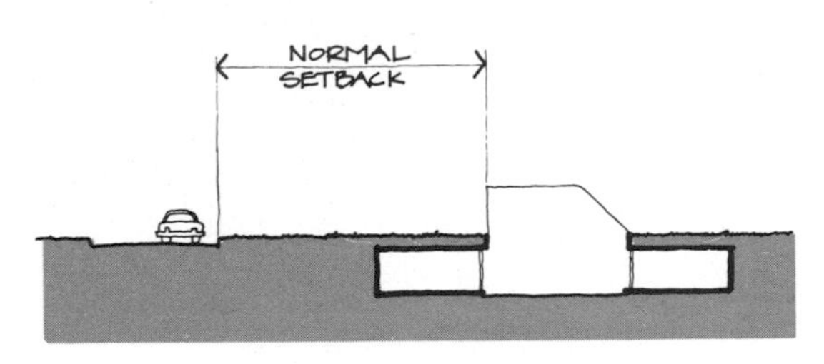

1-7 effects of lot size & adjacent structures

In an urban or suburban setting the size of a building lot and the proximity of structures on adjacent lots present some unique design considerations for earth sheltered housing. Three issues in particular arise if an earth sheltered structure is to be placed on a small site within a development of conventional above grade houses. The first is that some earth sheltered designs may require more manipulation of the land forms on the site and the creation of earth berms around the house. The additional area to achieve this type of design would not necessarily be substantial but may require a slightly larger site than normal.

The second issue concerns set-backs in a conventional development. A clear understanding of set-back limitations may be critical to some earth sheltered designs. It may be necessary to extend portions of a sub-surface structure beyond existing set-backs since arranging spaces around a courtyard is likely to take more area than a conventional rectangular house. A redefinition of zoning ordinances may be required in a case such as this.

A final consideration is the location and size of structures on adjacent property which may interfere with views, block sunlight or simply create an unpleasant feeling of being looked down upon in an earth sheltered house. Although these factors may present problems they can often be dealt with effectively by the designer.

It is important to note that these issues only arise when earth sheltered housing is used on smaller lots in existing developments. New developments which include only earth sheltered housing can be designed at fairly high densities without encountering the problems of scale and set-backs mentioned here. These and other legal and zoning issues are discussed in greater detail in the public policy section of the report.

soil and groundwater

Since earth sheltered housing usually requires a heavier structure and may be placed more deeply into the earth than a conventional house, consideration of soil type and groundwater conditions are particularly important to site selection. Determination of the soil type is mainly important for proper structural design of footings and walls. Certain types of soils can be unsuitable due to their poor bearing capacity or their tendency to expand when wet. Groundwater conditions are important to determine because of their impact on waterproofing as well as structural design. A high water table may require more costly structural and waterproofing techniques and make a site unsuitable.

Because of the serious consequences, the soil type and groundwater conditions of a potential site should be determined before any great effort or capital is expended on either the site or the design for the house. This presents a problem for someone who wants to build an earth sheltered house and is trying to find the right piece of land. If he or she has a site investigation done before the land is purchased, that investment may be lost if the site does not prove to be suitable. If no site investigation is done before the site is purchased, there is a possibility that the site may prove later to be unsuitable or else a very costly site on which to build. A compromise approach to this problem is, initially, find out as much as possible about local soil and groundwater conditions without actually having a physical site investigation carried out. Such general information will be more easy to obtain in more densely populated areas and sources of such information would include:

- Local soil exploration or soil testing firms
- Local consulting engineers
- City engineers or city offices
- Local realtors
- Owners of neighboring properties

With this preliminary information in hand, a more enlightened judgement can be made as to whether the site appears suitable (or at least that the apparent limitations are acceptable). If the results of the preliminary investigation are satisfactory a detailed site investigation should be done before the land purchase is finalized. The investigation may reveal some unforeseen problems that would make rejection of the site the best alternative. The main drawback to a site investigation at this stage, is the possibility that the owner of the land may not conclude the sale even though a substantial investment has been made. The cost of the investigation can, however, be divided between the two parties in a number of ways:

- The buyer and seller splitting the cost
- The buyer paying if the results indicate favorable conditions
- The seller paying if the results indicate unfavorable conditions

Generally, if the site is suitable from the other planning aspects, most types of soil will not greatly affect the design of the house. The groundwater and drainage characteristics of the site can have a larger impact on the design. For more detailed information on the effect of soil types and groundwater conditions on the design, as well as detailed site investigation procedures, consult the structural design section of the study.

2 architectural design aspects

programming

general design concerns

- impact of energy conservation
- building codes
- impact of structural systems
- relation to the surface
- typical layout considerations
- pedestrian & vehicular access

detailed design concerns

- natural lighting
- accoustics
- landscape concerns
- special energy considerations

introduction

The architectural design of any structure is developed from a broad range of criteria. These design determinants include everything from very general and subjective concepts to quite specific and technical details. In this section the intention is not to present a complete review of all the determinants which are relevant to housing design. Instead, the items discussed here represent design considerations which are unique to earth sheltered housing or have particular significance with respect to energy efficient design. These considerations are intended to aid prospective owners, builders, and designers in understanding and working with this type of house. The section is divided into three parts which include a discussion of programming followed by general and detailed design concerns. The general design concerns represent the major factors which influence the physical form of the design such as energy conservation issues, the structural system and the relation of the house to the surface. The detailed design concerns represent a number of pertinent details which have a significant impact on the design and energy efficiency of an earth sheltered house such as natural lighting techniques, and landscape concerns.

programming

Programming is the process of establishing the needs and desires of the users of a building and then reflecting these in the spaces to be designed and built. This means simply that the type and number of rooms, their size, and the relationship between the spaces must be determined before proceeding with the design. Since most families have similar basic needs such as eating, cooking, sleeping and recreation, the programs for most single family dwellings are basically the same. This similarity in home programs is reinforced by financial restrictions and the desire for conformity so that any home appeals to a broad market. Naturally, an earth sheltered home is subject to the same program determinants and market forces as an above grade home and must be designed to include the typical functions and space requirements a homeowner would expect in above grade construction. However, earth sheltered housing presents some unique opportunities as well as physical limitations which appear to have a significant impact on programming. These issues are discussed in the following paragraphs.

One effect of building a home either partially or completely recessed into the ground is that the large unprogrammed basement area found in a typical above grade house will probably be more costly or impractical to duplicate. The basement is considered to be relatively cheap space in above grade construction in Minnesota since foundation walls and footings are required to be at least 107 cm (42 in) deep to avoid frost heave. In an earth sheltered design, the walls are likely to go below that depth so the opportunity for cheap extra space is not present. The typical basement does serve several important functions to a conventional house which include:

- Laundry room
- Mechanical space
- Storage area
- Workshop area
- Recreational or multi-purpose space

These functions must be recognized as needs and be included in the program for an earth sheltered home. It is unlikely that they will require the extensive left over space which they often occupy in a typical basement. It is possible that a more efficient living arrangement and use of space may result by combining these basement functions into the rest of the house.

One of the functions mentioned above which deserves special attention in the program is the mechanical equipment space. Based on data presented elsewhere in the study, it appears that the heating and cooling capacities of the equipment in an earth sheltered house will be much smaller than equipment for an above grade house of equal size. However, the physical size of the equipment may not decrease noticeably. It is desirable from a noise and safety point of view to locate the equipment within a mechanical room. The equipment alone will occupy about 5.6 sq m (6O sq ft) based on the following assumptions:

- Water softener required
- Solar domestic hot water preheat utilized
- Furnace is natural gas or electric
- Dehumidification or air conditioning utilized
- All ducts are ceiling mounted
- Controls are wall mounted
- No grey water heat recovery capability

An additional 2.3 sq m (25 sq ft) may be required for a circulation aisle. Greater efficiency of space utilization can be achieved by combining the mechanical room and laundry area in such a way that the circulation is shared.

A related program consideration for earth sheltered housing is the potential use of alternative mechanical systems such as solar collectors, ice air conditioning or heat recovery from waste water. The use of a solar collector system for heating is quite likely in earth sheltered designs since it compliments the reduced heating requirements and the thermal mass characteristics. In any design employing an active solar collection system additional space for equipment and storage capacity will be required. Water and rock are the two most likely storage mediums. A solar collector also may have a significant impact on the size orientation, and overall design of the house if it is attached to the structure. Another example of an alternative mechanical system which affects the programming and design of a structure is an ice air conditioning system which is presently being developed. In this type of system a separate underground chamber is placed near the house for the freezing of ice in the winter which is tapped for cooling in the summer. Such a system would have implications for additional equipment space as well as the siting of the house in relation to the ice storage chamber.

These and other alternative systems appropriate for earth sheltered housing are discussed further in the energy use section. The many possible alternative energy systems that can be utilized in a home all have effects on the design which are often significant. For this reason, a careful assessment of these systems during the programming phase will allow them to be properly integrated into the design.

A final program consideration that arises with earth sheltered housing is the potential use of unheated above ground spaces for certain functions or during certain times of the year. Since spaces such as a garage or a storage area do not have the same heating and cooling requirements as a habitable space, it may prove to be more convenient or less costly in some cases to locate them above ground adjacent to the earth covered portion of the house. Other spaces which have seasonal uses, such as a porch, may also be located above grade which could enhance their use. This issue is mentioned because an earth sheltered structure represents a distinctly different type of construction than that of an above grade structure and the fact that certain functions could easily be located above grade without reducing energy efficiency should be recognized in the programming stage of the design.

impact of energy conservation

One of the major reasons for designing and building earth sheltered housing is the potential energy conservation to be derived from it. Details concerning amounts of earth cover and specific projected energy savings are presented in the energy use section of the report. However, energy conservation is also discussed here since it is a major determinant of the architectural design. There are two ways in which energy conservation directly affects the overall configuration of an earth sheltered house. These are the development of a compact plan geometry and the maximization of the earth mass around the structure.

The heat loss, and thus, the energy use of any building is a function of the area of the surface through which the heat can escape. A building with a larger surface area will experience greater heat loss than one with a smaller surface area assuming all other variables are equal. This seems obvious, however, it must be recognized that buildings with the same floor area can have quite different surface areas depending on the configuration of the plan, as shown in the adjacent illustrations. Assuming a one level structure with a flat roof, the shape which encloses the most space with the least wall area is a circle. Since a circle is often impractical to construct, a square or rectangle represent relatively compact plan shapes for building design. Similarly, when one level and two level designs with equal floor areas are compared, the two level design has far less exterior surface area. The more extended and less compact a design becomes, the greater the surface area exposed to heat loss. This principle holds true for earth sheltered housing although it should be mentioned that this is not a detriment in the summer. Since the surrounding earth is cooler than the house in the summer, the flow of heat into the earth is major source of cooling. In this case a house with a greater wall area will benefit more from this cooling effect. Generally, in the Minnesota climate, more emphasis is likely to be placed on mimimizing heat loss in the winter and therefore, developing as compact a floor plan as possible.

The second major way in which energy conservation affects the overall design of an earth sheltered house is related to the earth mass surrounding the structure. As will be shown in the energy use section of the report, earth placed against the walls and on the roof of a structure have definite energy saving advantages. Therefore, the maximization of earth cover becomes a primary determinant of

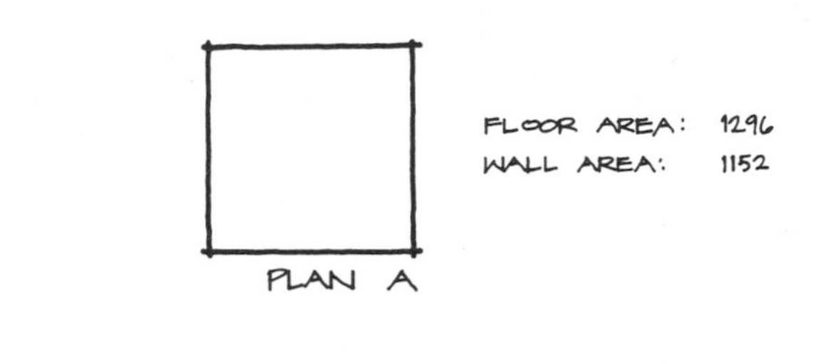

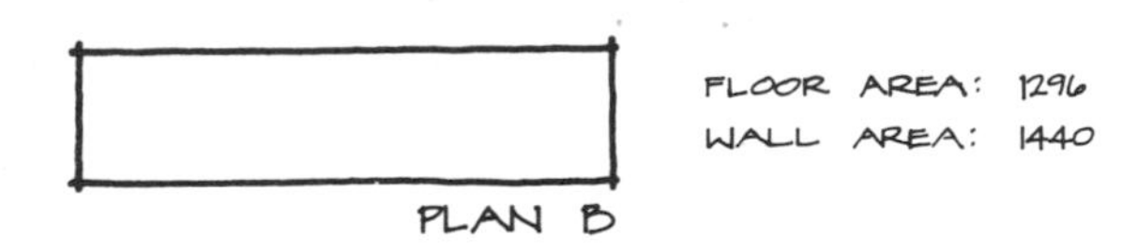

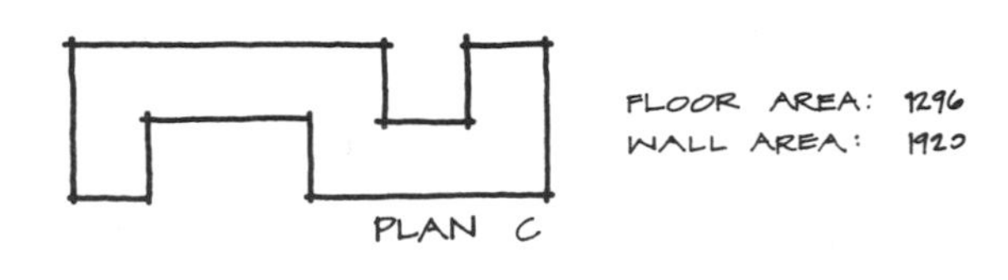

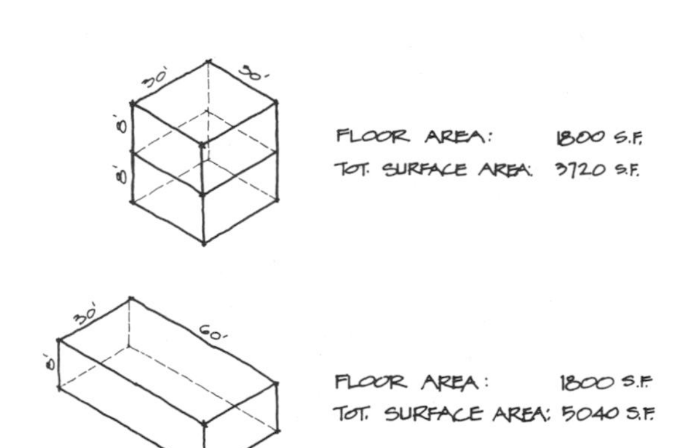

2-1 floor area vs. surface area

the design. From an energy conservation point of view alone, the ideal design would be a totally enclosed chamber well below the surface. Naturally, this would be an unacceptable living environment as well as being in violation of building code restrictions. Window openings, courtyards, skylights and other techniques that are required to create a livable environment can be designed without totally eliminating the significant energy savings the earth covering offers and can actually enhance the energy savings in some cases.

building codes

The implications of building codes for earth sheltered housing and recommendations for future policy are discussed in a later section of this study. However, it is important to briefly mention here one of the most profound issues that arises related to building codes when applied to this type of housing design. For residential construction the building code states that all habitable rooms must have an operating window to provide light, ventilation, and a means of escape. Rooms such as storage and utility are excluded and bathrooms are excluded provided mechanical ventilation is present. In typical above grade construction these provisions can be met without significantly affecting the design. However, unlike a typical house, many earth sheltered designs tend to minimize and concentrate window openings in order to maximize energy efficiency. If this code requirement is strictly adhered to it becomes one of the most significant design determinants since it may require unwanted additional windows or force unusual plan arrangements so that windows can be grouped together. There is no question that an earth sheltered home without adequate light, ventiliation, safety standards, and a good relationship to the outdoors is totally unacceptable. However, modification of this particular window requirement allowing for other techniques of bringing light and air into spaces would have a very liberating impact. Generally, the design considerations and designs presented in this and other sections of the study will be done within the limitations of present building codes. In some cases modifications of code requirements will be illustrated and noted as such.

impact of structural systems

In any building, the structural system is an important shaping force of the overall design. This is particularly true with earth sheltered housing since the loads resulting from earth on the roof are substantial. The structural systems that can be used to support these loads can be divided into two groups: the more conventional flat roof systems and a variety of more unconventional systems using vault and dome shapes.

The conventional roof systems mentioned above include precast concrete planks, poured in place concrete slabs, as well as wood or steel post and beam systems. All of these systems have common general characteristics which result in similar design configurations. These characteristics result in flat or sloping roofs, and generally rectangular plan shapes.

The necessity of supporting heavier than normal roof loads in earth sheltered housing may result in the use of the more unconventional structural systems mentioned above. These include concrete or steel culvert shapes and geodesic domes. These systems can support the heavier loads in a much more efficient manner than flat roof structures, although they can limit the alternatives for room layouts considerably. More than the conventional systems, these structures dictate the overall shape and form of the house and the spaces that must be placed within these shapes. Also, systems depending on arch or vault action for structural support cannot be penetrated for window openings without regard for the weakening of the structural system.

For example, if a large culvert shape is used as a basic structure, window openings are primarily limited to the two ends and the spaces must be laid out with regard to the curved roof. Possible alternative design approaches may include a shell shape large enough for two floor levels or several smaller shapes linked together to create more openings. Plans and photographs of an existing earth sheltered house utilizing steel culvert sections are included in Part B which also contains a conceptual plan for an earth sheltered geodesic dome.

These types of systems can result in very attractive designs with some very unique spaces while maximizing the effect of the earth mass. It must be recognized, however, that most unconventional systems will present a different set of limitations on the design and ultimate configuration of a house.

Most of the discussions and illustrations pertaining to various aspects of earth sheltered housing throughout this report refer to designs using conventional structural systems. At this point, it appears that the vast majority of single family dwellings will be constructed with conventional systems. However, it should be noted that all of the design and energy considerations discussed in the report apply in principle to earth sheltered housing with any structural system, whether conventional flat roof or unconventional shell structure.

relation to the surface

One of the key considerations in the design of an earth sheltered residence is the relationship of the structure to the surface. Generally the relationship to the surface refers to the actual depth of the structure and the manner in which it is placed on or into the earth. On a flat site, an earth sheltered home can be fully-recessed into the earth or partially recessed with earth berms formed against the outside walls. On a sloping site, a structure can be set into the hillside in various manners. For example, on a slightly sloping site, the design may be fully or semi-recessed similar to a flat site. On a more steeply sloping site the opportunity for a two level design set into the hillside presents itself. The depth of the structure and thus the surface relationship is also affected by the amount of earth placed on the roof. Some designs which rely principally on earth covered walls for energy saving may have no earth cover on the roof at all.

There are two major design issues associated with the relationship to the surface. The first is the view that can be seen from inside the house. If a site has a desirable scenic view or if the inhabitants simply want a more open relation to the outdoors, then a semi-recessed design offers a better alternative than a fully recessed design on a flat site. The view is even more open and clear with a house recessed into the hillside of a sloping site. There are other design alternatives which may solve the conflict of an open view versus placing the structure below grade to save energy. One alternative is a two level design which places a minimum amount of space such as a living room or den above grade with the rest of the house below grade. Although this offers a clear open view, the energy trade-offs may be significant. It should be noted that many sites do not have desirable exterior views and that placing the rooms around a courtyard may provide an excellent alternative to a typical above grade yard space. There

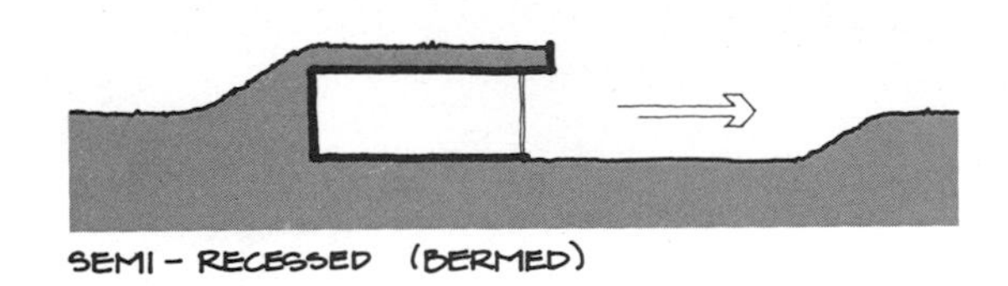

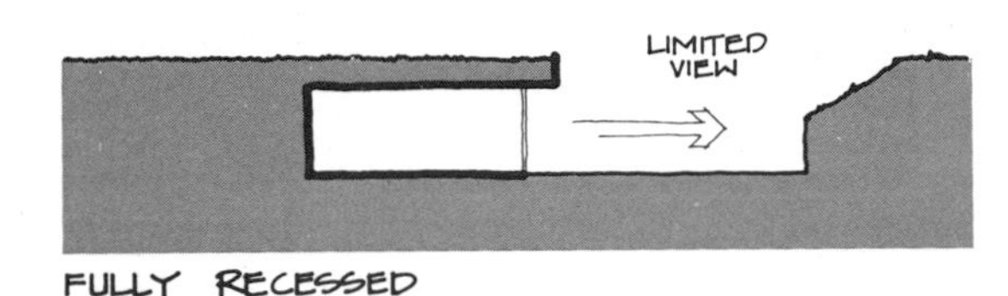

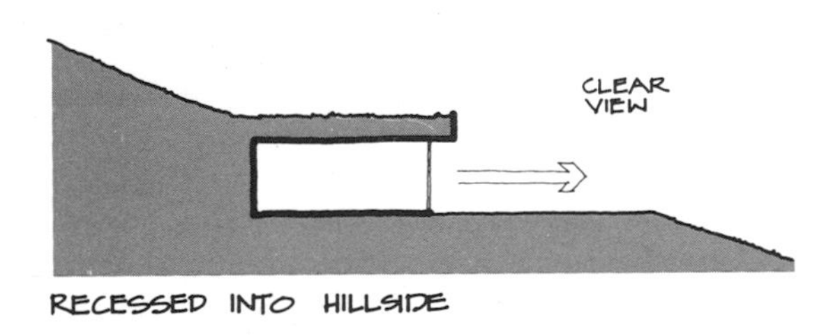

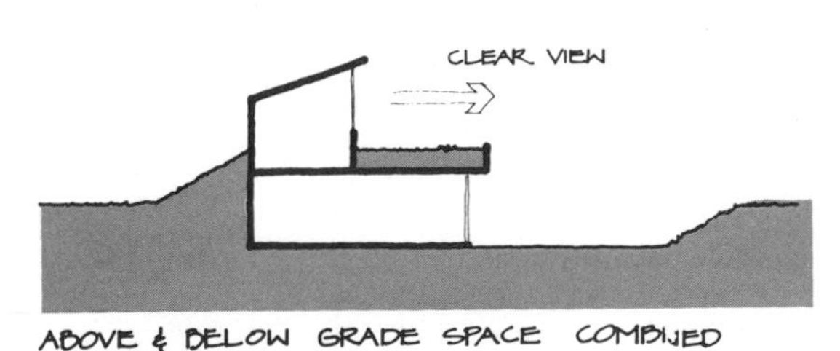

2-2 relation to surface

can be a greater sense of privacy and enclosure with an atrium type of plan, and the relationship to the surrounding earth may be less significant than with other designs relying on an outward view.

The second design issue associated with the relationship of the house to the surface is simply the overall form created by the house. The form of the house has certain implications in the appearance, implied privacy and security, and the use of exterior spaces. For example, on a flat site, a semi-recessed house would have a different appearance and feeling from the outside than a fully-recessed house. The earth berms around the semi-recessed house create more of a visual barrier to outsiders and may be used to define some exterior spaces around the house. Also, the bermed type of house has the advantages of diverting surface water runoff around the structure as well as raising the floor level higher above the water table.

Another design alternative which affects the overall form is the placement of a garage and entry or a living space above grade with the remaining spaces below grade. The above grade portion becomes the predominant visual form of the entire house and it can separate and define exterior spaces. The space on the roof of an earth covered structure is extra space which typical homes do not have and may be more accessible and private when entered from an adjacent above grade space. Naturally, the form of an earth sheltered house and its implications varies with each individual design and can only be discussed in general terms here. Nevertheless, it is important not to overlook some of these more subjective considerations such as appearance and feelings of enclosure since they will greatly affect the success of the entire design.

typical layout considerations

In order to discuss various issues which are related to the actual layout of earth sheltered housing it is convenient to define three basic plan concepts which demonstrate unique characteristics. These are the elevational, the atrium and the penetrational types. The difference between these three concepts is basically in the size and orientation of the window openings. The elevational type concentrates all openings on one exposed elevation with the remaining three sides earth covered. The atrium type places all openings around a central courtyard with earth surrounding the outside of the house. The penetrational type uses window openings of various sizes which penetrate the earth covering in various locations around the perimeter.

These three distinct concepts in window massing all have different implications for room arrangements and internal circulation which are discussed in the following paragraphs. It should be noted that these three concepts simply represent general categories of plan types and that many combinations and variations can be developed such as two level schemes or designs combining above and below grade spaces.

elevational

The general concept of the elevational plan is to maximize the earth cover around the house by concentrating all of the window openings on one side of the house, preferably the south side. The plan then must be arranged so that all the major living and sleeping spaces are along the exposed elevation with secondary spaces not requiring windows (such as baths, utility and storage) placed behind them. A kitchen or dining space may be placed away from the windows if it is part of the general living space in front of it. This arrangement is similar to a typical apartment plan with one window wall. The disadvantage of such a plan is that the internal circulation can become rather lengthy, especially for a larger house, since all of the spaces are essentially lined up like the rooms in a motel. If the opportunity exists for a two level elevational design, a more compact plan can be developed without lengthy circulation. Based on information presented in the energy use section of the report, it also appears that there are definite energy saving advantages with a two level design. Due to the continuous earth mass, the placement of all windows on the south, and the relatively compact layout, the elevational concept is generally a very energy efficient one and is used for energy use projections and comparisons later in the report.

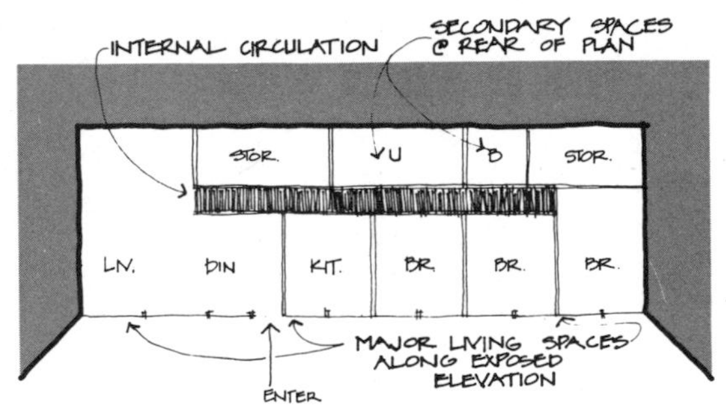

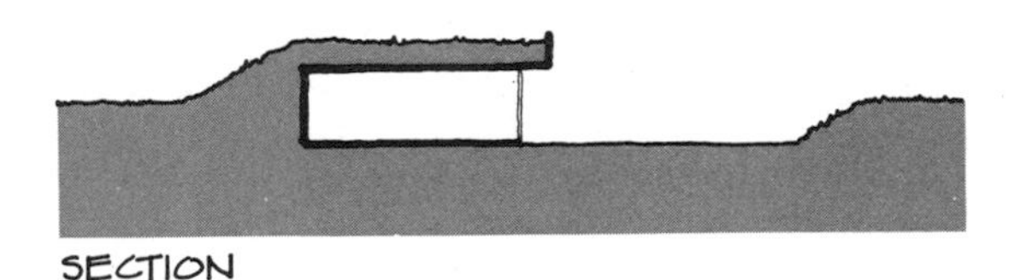

2-3 elevational plan

atrium

The atrium plan is based on the idea of placing the living spaces around a courtyard with all the window openings oriented into the court. In its simplest form an atrium is a square court with spaces on four sides although it may be desirable to place the living spaces on just three sides leaving one side open for

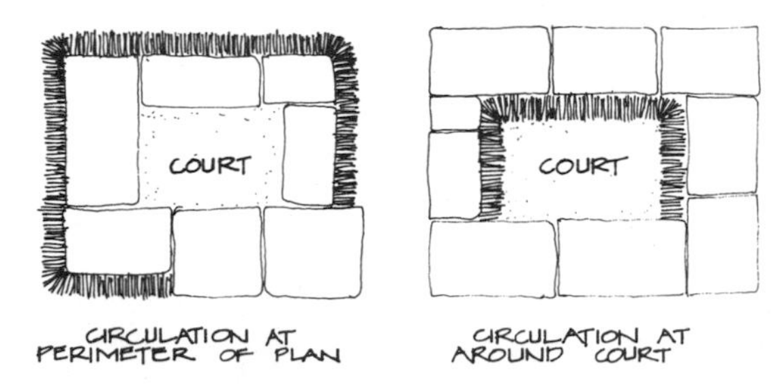

2-5 internal atrium circulation

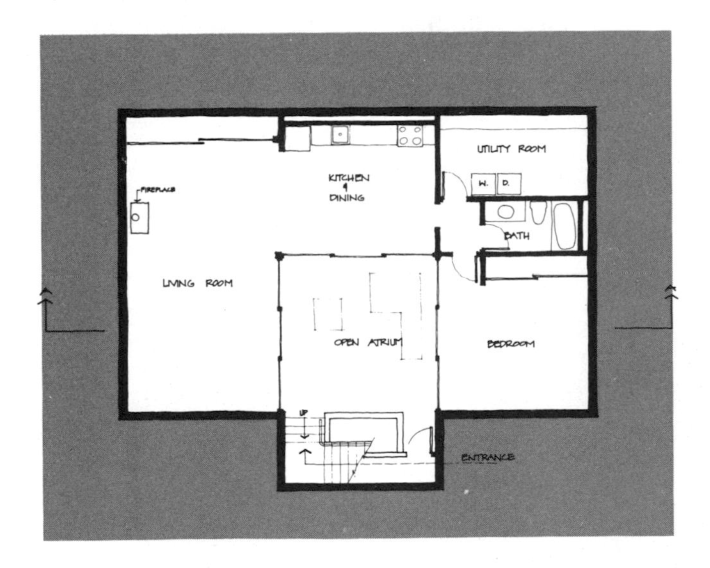

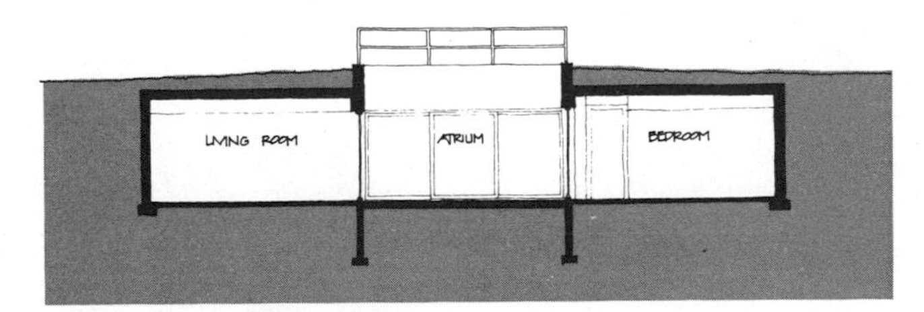

2-6 one bedroom atrium

light, view and access. Although the general concept of an atrium is a promising one, there are difficulties with internal circulation that must be solved. Traditionally, the atrium plan was developed in warm climates and the courtyard itself was used for circulation. Some earth sheltered houses have been developed along these lines in warmer parts of the United States such as the Bordie residence in Austin, Texas, shown in Part B of the report. In Minnesota and most of the United States, it is necessary to reach all spaces in the house without going outside. One simple alternative is to cover the atrium and create an interior courtyard surrounded by the living spaces. In this case, however, the spaces do not open directly to the outdoors which may require a code variance. Internal circulation to spaces around a courtyard can be provided around the perimeter of the plan so that all spaces have a clear view to the courtyard, but this results in circulation corridors which are are too long and inconvenient to be justified in a residence. The other alternative is circulation around the court itself which reduces the corridor length to an acceptable level. The circulation must now, however, pass between the spaces and the courtyard. It is acceptable to pass through open spaces such as a living, dining or even kitchen area in this manner but private spaces such as bedrooms cannot be used as corridors nor can they be cut off from windows without adjusting present building codes.

This problem of internal circulation in an atrium type of plan is basically a factor of size and building code requirements for windows. A single atrium is a valid plan concept up to a limited size such as the one bedroom plan shown here where circulation only occurs through the dining area. With a larger program it becomes necessary to modify the simple atrium idea in order to provide a compact, efficient plan with adequate window openings to all the required spaces. Some alternatives are the use of two or more atriums, additional window openings through the earth berms, or a two level design. Another alternative is a modification of present building codes so that an outside window is not required in a bedroom as long as adequate ventilation, light and means of escape are provided.

Although the atrium type of plan does not face all windows to the south and is not as compact as an elevational plan the courtyard does tend to trap air which is then heated by the sun, thus reducing the heat loss somewhat. It also has some attractive features such as the enclosure of a private outdoor space that make it a valid concept for earth sheltered housing.

penetrational

The penetrational type of plan places window openings of various sizes in various locations around the house. An opening in the earth berm occurs at the window locations which allows for light and view from any space on the perimeter of the structure. Basically, the penetrational concept is no different from a conventional above grade house in that windows can occur on all sides and the plan arrangement can be quite compact with circulation limited to the center of the house. However, in an earth sheltered design where energy saving is a primary consideration, the location, size and grouping of window openings must be carefully considered. Although a penetrational plan is not restricted to placing all windows on one wall as with the elevational schemes, it is still desirable to place the major living spaces with the largest window openings on the south side and the less active spaces with as few windows as possible to the north. It is also desirable to simplify the plan so that the maximum amount of earth cover can occur without interruption. A few larger openings are likely to be easier to construct and save more energy than many small openings where the earth mass is constantly penetrated and cannot act as a large thermal mass. On many sites it may not be desirable or even possible to face all windows to the south (as in the elevational plan) thus a penetrational layout becomes a reasonable alternative.

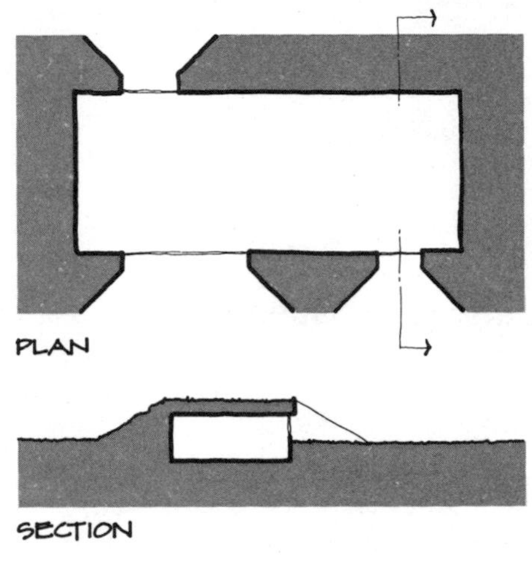

2-7 penetrational plan

pedestrian and vehicular access

One very important concern with earth sheltered design is the pedestrian and vehicular access or means of entry to the house. The entry is often a focal point for building designs but it has a particular significance with earth sheltered structures since some people have negative associations with going underground and the entry can alleviate these misconceptions. An entry should be obvious to a visitor from the outside, and once entered should be light, spacious and not require an excessive number of steps to arrive at the main living level. Referring to the plan types mentioned in the previous section, the elevational plan can be entered most simply on the open side, preferrably near the center of the house for more efficient circulation. However, it is often desirable to separate the formal public entry to the house from the more private exterior spaces which would occur along the exposed elevation. In this case, an entry could be placed along any of the remaining three sides of the house by interrupting the earth berm although entering from the ends would probably result in inconvenient and excessive circulation inside the house. In a similar sense, the simplest point of entry with an atrium plan is through the courtyard itself. This solution again presents the conflict that a courtyard used for public entry may not have the desired privacy and seclusion. An atrium scheme could be entered through a separate entrance on top of or through the surrounding earth berms. In a plan with two courtyards, one could be an entry space and the other a more private space. In a penetrational plan, the entry can occur much as it would in a typical above grade house. As with all earth sheltered designs the entry must be carefully designed to provide a pleasing transition to the level of the house .

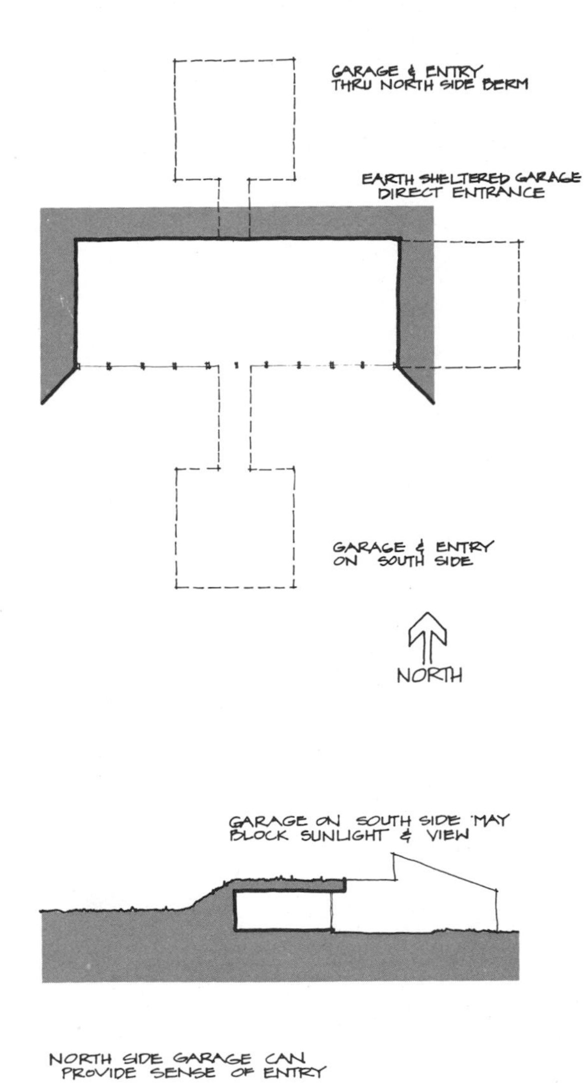

An entry to a house is usually related to the vehicular access and garage. In an earth sheltered design the garage could be made part of the earth covered structure or it could be placed above grade perhaps connected to the entry. This presents some unique problems and opportunities in design. An earth sheltered garage may blend into the overall design better but the cost may be higher than a typical garage. An above grade garage placed on the exposed side of an elevational plan may affect the sunlight and view. However, on the opposite side of the house it may be combined with an entry to create a visible form and separate the public and private sides of the house. The designs shown in Part B illustrate a variety of ways to deal with the unique problem of integrating a garage with an earth sheltered house.

Concern has been expressed that earth sheltered housing may be inaccessible to handicapped persons. This is based on the incorrect assumption that this type of space is always completely below grade and requires extensive stairways for entry. Many designs presented in this study illustrate the fact that earth sheltered housing can often be easily entered from the existing grade and presents limitations for handicapped persons which are no different than a typical above grade house. Designers should be aware that only slight alterations in certain details of steps and entrys can make a great deal of difference in handicapped accessibility in any structure.

detailed design concerns

natural lighting

In earth sheltered housing the penetration of natural light into the living spaces is important for two reasons. First it is desirable in any design concerned with energy efficiency to allow as much solar radiation to reach the interior as possible, particularly since passive solar heat is basically free energy. Second, earth sheltered design is sometimes incorrectly associated with the darkness of a basement and it is important to allow sunlight to enter the spaces simply to create a brighter, more livable environment. It is necessary to consider the position of the sun at various times of the day throughout the year in order to properly take advantage of the light. The adjacent illustration indicates the altitude of the sun during the year as it penetrates a south facing wall at noon. An overhang can be used to keep the summer sun out while still allowing the winter sun to enter. A trellis can be used as an overhang which shades the south elevation in the summer and as the leaves drop off in winter the sun can penetrate into the living spaces more fully.

One problem which should be considered with a design that concentrates all of the windows on one side (elevational) is that there is a limit to the size of space that can be adequately lighted from one wall. A typical room up to l6 ft deep can be lighted from one side but a larger space with a kitchen or dining space away from the windows may require additional sources of natural light. Several techniques could be considered to introduce light to a space which is away from the major window opening. These include skylights, a sloping roof allowing greater light penetration from the window wall or simply additional window openings penetrating through the earth berms if necessary. Of course, all of these design alternatives have cost and energy use implications.

A skylight, for example, can represent a great energy loss if a typical flat or bubble type of skylight is used. However, certain designs can be far more energy efficient. A directional skylight facing south, similar to the one shown in the illustration, can be designed so that passive solar radiation is reflected into the spaces in the winter but is screened out in the summer as the sun angle changes. This type of skylight can also be designed with operating windows and used very effectively for ventilation in the summer. The energy loss though any skylight can be greatly diminished by using some type of insulated shutter over

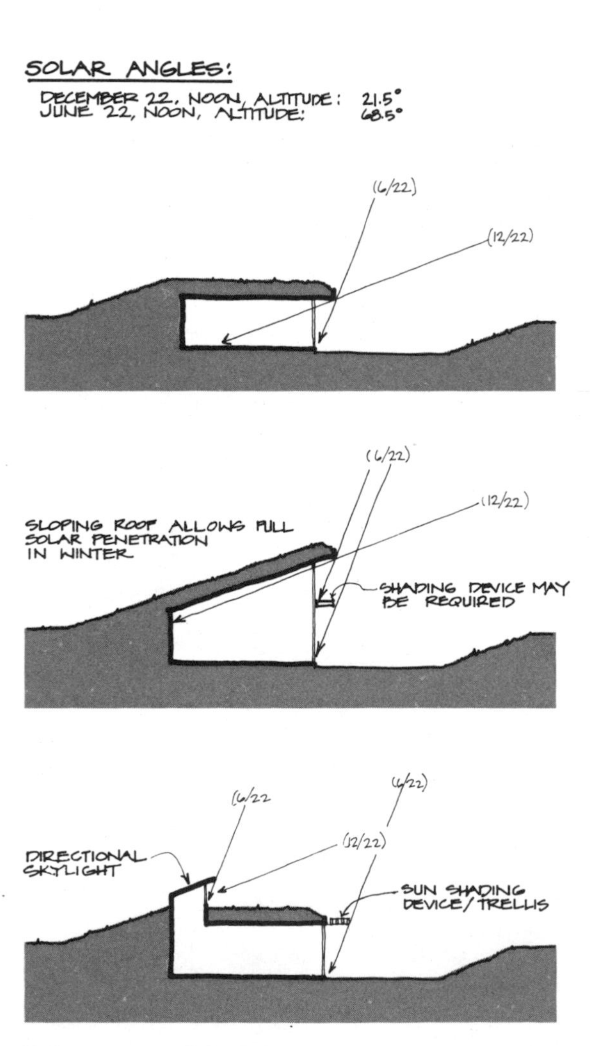

2-9 natural lighting

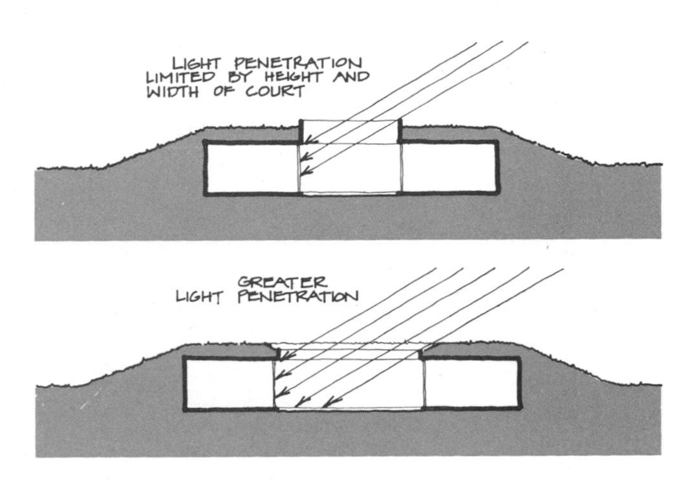

2-10 atrium natural lighting

the opening at night. References to additional information on skylights are given in the bibliography.

A final lighting consideration should be mentioned which pertains to atrium types of plans. The length and width of a courtyard and the height of the surrounding structure will have a significant effect on the amount of light which reaches the courtyard and the adjacent living spaces.

acoustics

One effect of building an earth sheltered house that deserves mention is the great dampening of outside sound and vibration by the surrounding earth mass. In fact, one reason a company which manufactures precision instruments in Kansas City decided to locate their operations underground was to eliminate vibrations from the outside. In housing, this acoustical isolation is a definite benefit on sites which are close to busy highways, airports, or other undesirable noise sources. It should be noted that the greater degree of quiet in an earth sheltered house may actually make common noises such as appliances or mechanical systems seem louder in contrast to the silence. This effect may have to be dealt with by dampening these normally unnoticed sounds, or providing a slight amount of background noise.

landscape concerns

In an earth sheltered design, landscaping cannot be considered as a separate decorative feature to be added after the house is built. It is a critical component of the overall design which must be coordinated with all of the other elements of the house, particularly the structural and waterproofing systems. This section presents the major landscape concerns which are unique to this type of housing.

A very complex and critical landscaping problem is that of an earth covered roof which is part of many earth sheltered designs. It is desirable to have plant growth on the roof not only for aesthetic and ecological reasons but for energy savings as well. It is shown in a later section of the study that the reflective nature of grass and groundcovers help reduce solar heat gain considerably in summer. The important factors in the design of an earth covered roof are the depth of soil, type of soil, and the method required for proper drainage. The energy efficiency of the house and the potential for plant growth are both enhanced by the greatest possible depth of soil. However, the weight of soil is considerable and the structural system required to support a few feet of earth is usually costly for a house. Therefore, it is necessary to know the minimum amount of soil required for various types of plant materials. The illustration, which is taken from a recent article by Thomas E. Wirth, (see bibliography), indicates that 30 cm (l2 in) to 46 cm (l8 in) of soil is adequate for grass and other groundcovers and 6l cm (24 in) to 76 cm (30 in) for small shrubs. Larger shrubs and trees require soil depths up to l52 cm (6O in) which is likely to be too much to support at a reasonable cost.

Information concerning the exact depth required for a certain plant should be verified by a landscape architect since variations in plant type, soil type, moisture and climate all will affect plant growth. In larger commerical structures and public buildings with earth covered roofs or rooftop plazas, techniques have been developed for reducing the total soil weight but still providing tree wells by using styrofoam to displace large volumes of earth. These techniques seem inappropriate and unnecessary for most housing since the size of the roof is smaller and large trees can be located adjacent to the structure much more easily and inexpensively than on top of it.

It is essential to provide proper drainage for soil in a rooftop planting situation. Plants will not survive if the soil is completely saturated. It is helpful to provide some slope to the earth on the surface to divert excessive moisture off the roof. It is also important to provide a drainage layer beneath the soil so that the water which filters through the soil is carried away. This is most often done by providing a gravel layer with a soil separator matting to prevent the soil from settling into the gravel and hindering drainage. The previous illustration indicates that the gravel is placed over the waterproofing and insulation which are usually on a slightly sloping roof which aids drainage. There are many variations on this detail but the principle of providing drainage beneath the soil is usually the same.

There are other aspects of landscaping an earth covered roof which deserve mention. One is that it may be necessary to provide barriers along the edges to prevent people from falling into a court below. Using small shrubs to protect the edges may be more attractive than a fence or they may be used to hide a fence if one is required. Another aspect of roof top planting which is quite unique

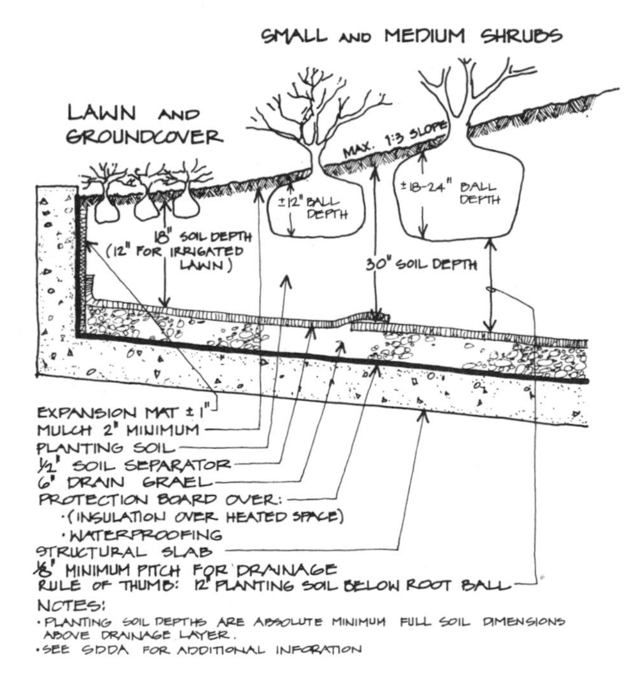

2-11 planting details

compared with other types of landscaping is the effect of the heated space below on the earth layer and plants above. The growing season and the actual plant growth may be modified by the thermal effects of the building. The impact of this is difficult to determine and little research appears to have been done, but some unpredictable results may occur.

Other aspects of landscaping an earth sheltered house are much more conventional than the roof. However, they should definitely not be overlooked or treated as an afterthought. The final appearance and success of this type of house will depend heavily on how well it is integrated with its site. To a greater extent than with a conventional above grade house, landscaping is an essential element in the design since the house is really part of the surrounding landscape rather than simply placed on top of it.

special energy considerations

In designing an earth sheltered house, there are many details which must be carefully analyzed because of the unique conditions of placing a structure completely below grade. Also, many conventional construction details are not acceptable because they detract from the energy efficiency of the structure. In this section of the study, some special energy considerations are presented.

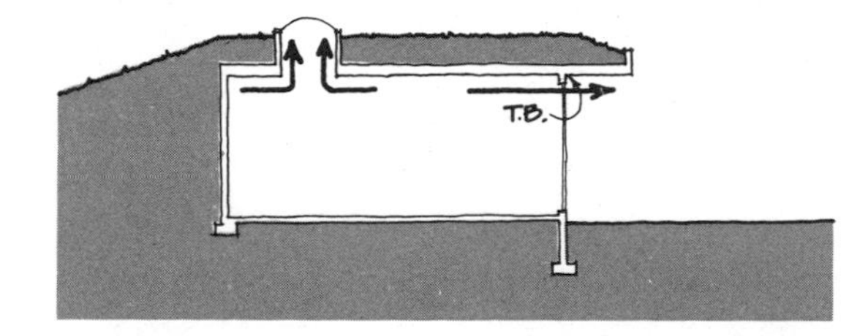

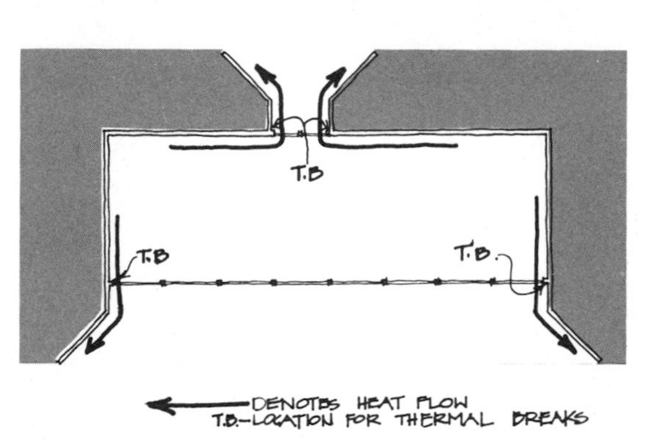

2-12 thermal breaks

thermal breaks

Since many earth sheltered structures are likely to be constructed of concrete, the problem of losing heat by conduction through the concrete is a major one. This occurs when a concrete roof or wall is continuous from the inside to the outside of the house in order to form an overhang or a retaining wall. This also occurs with skylight wells, parapet walls or any other situation where the concrete structure acts like a wick which allows the inside heat to bleed away to the outside. This problem can be dealt with in a number of ways. The simplest is the elimination of these details from the design, but this is often not practical considering the necessity of retaining the surrounding earth forms. Another solution is providing a thermal break in the concrete such as a layer of styrofoam insulation which effectively separates the interior and exterior structures. This will cause the two parts of the structure to move and act separately which must be considered in the structural design. The use of this type of thermal break is illustrated by the Jones house in Part B. A third solution is the use of a separate structure and material for the exterior parts. This would include, for example, the separate trellis structure suggested previously for solar shading of windows. Also, retaining walls around window and door openings could be constructed of wood timbers which would be separate from the concrete structure and solve the heat loss problem. Naturally, these transitions from indoor to outdoor must be integrated with the structural and waterproofing systems as well as the desired appearance of the house.

covered atrium

In the atrium type of design where the interior spaces look onto an exterior courtyard, there is a unique opportunity to increase the energy savings of the house. In the winter, a temporary transparent cover of glass or plastic could be placed over the courtyard thus enclosing it as an interior space. This type of space could provide a great deal of passive solar energy. The windows of the house could be opened to receive the heat from the atrium in the day and closed to it at night. It should be noted that this idea may be in conflict with code requirements that state that habitable spaces must have windows opening directly to the outside. One example of this concept is illustrated in Part B of the report, so it is obvious that the code conflict can be resolved in certain situations. This is clearly an opportunity to enhance the energy efficiency of the house and create a pleasant additional indoor space in the winter although the cost of such an addition must be carefully considered. It is advantageous to consider providing such a cover early in the design phase of the house in order to integrate it with the overall design even if it will be an option that can be added later.

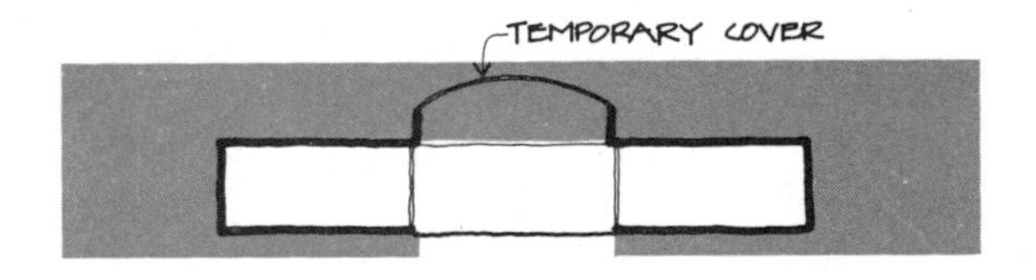

2-13 atrium cover

thermal shutters

Another design detail which can have significant energy saving impact is the use of insulating shutters or drapes over the windows. This particular idea is obviously applicable to all types of housing concerned with energy conservation and is not restricted only to earth sheltered structures. There are numerous designs for insulating shutters that are applied to either the inside or the outside of the windows. One example is a flexible shutter on tracks which rolls down over the outside of the window and is operated from the inside by a motorized system.

Another is a shutter on the inside made of styrofoam covered with wood or cloth which opens into the space like a bi-fold door. The basic principle of all shutters is simply to create a far better insulated wall when the window is not being used to collect passive solar energy. This is usually done by opening the shutters at dawn and closing them at dusk although additional energy savings will occur if the shutters are closed on cloudy days or anytime the sun is not directly shining into the house. Draperies are used in the same manner as shutters although they are unlikely to have as great an insulating capacity or to be sealed as tightly as a well-designed shutter. For maximum effect, drapes should be heavy, preferably insulated, and should fit tightly to the wall around the window to prevent convection currents flowing across the window surface from entering the room.

3 energy use

- **thermal characteristics of earth sheltering**
  - roof
  - walls
  - floors
  - overall considerations
- **energy efficient design concerns**
  - heating sources
  - cooling sources
  - method of heating/cooling delivery
  - ventilation
  - humidity controls
  - hvac controls
- **energy analysis**

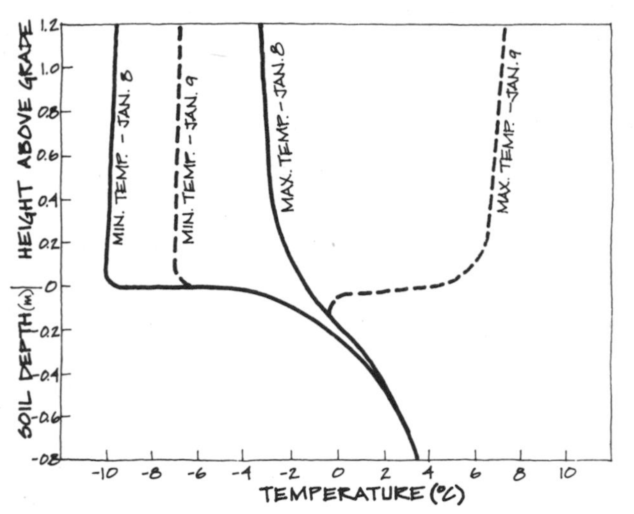

3-1 tautochrone

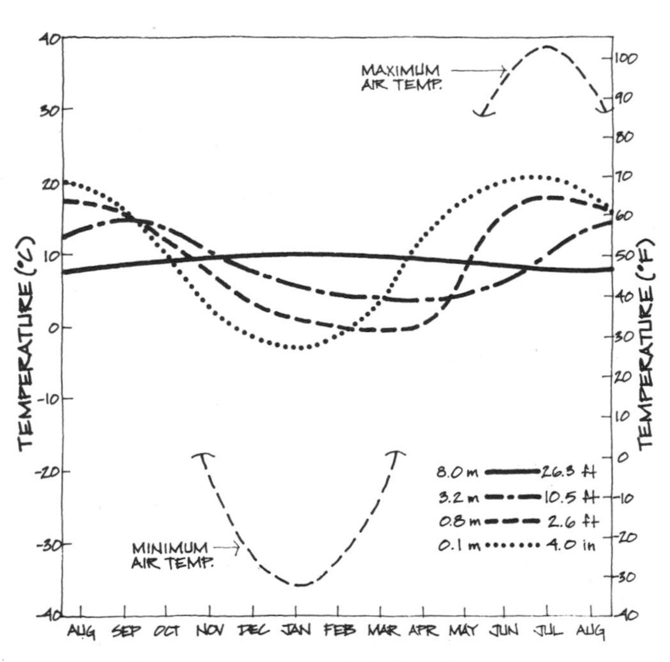

3-2 soil temperature distribution

The great recent interest in earth sheltered housing has resulted primarily from anticipation of substantial energy savings. In typical above ground construction energy is wasted by unwanted heating or cooling of the surroundings. By reducing heat transfer to and from the surroundings less energy is required to maintain desired conditions. Heat loss (or gain) from a structure principally depends on two factors: the ventilation load for heating or cooling intake air, and the heat transmission through the building envelope. In most residences the ventilation load consists of uncontrolled air infiltrating through cracks and holes. These infiltration losses are reduced greatly or eliminated by earth covered construction. Ventilation air then can be controlled so that heat recovery systems can be used effectively. The transmission losses depend on the amount of heat conducted through the envelope of the structure. This is a function of the thermal transmission coefficient, which is reduced by adding insulation, and the temperature difference between the inside and outside of the wall.

Above ground the temperature difference is determined by the local weather conditions. On the other hand, the earth averages the temperature fluctuations both on a daily and yearly basis. Seasonal temperature fluctuations reach a depth of several meters into the soil whereas the penetration of short term temperature fluctuations over periods of hours or days is almost negligible. The short term fluctuations are illustrated in the tautochrone (Fig. 3-1). The wide daily air temperature swing is essentially eliminated below 0.2m; this demonstrates the advantage of even 0.2m of earth cover. At greater depths soil temperatures respond only to seasonal changes and the change occurs after a considerable delay. As shown for the Minneapolis-St. Paul area in Fig. 3.2, the amplitude of the mean temperature fluctuation decreases rapidly with depth. At 5 to 8 meters the temperature is almost a constant 10°C (50°F), a value which is 3 to 4°C above the average yearly temperature for this location, and only 10°C (18°F) less than comfortable room temperature.

This temperature is above the yearly average since during the summer, heat is transferred down into the soil by warm water percolation and by conduction which is increased since the soil is wet. During the winter however, the surface freezes preventing heat transport by water. The soil dries lowering the heat conduction, and in addition, the land surface is covered by an insulating blanket of snow. Observe the rapid increase in temperature during April, May, June and the slow decrease in September, October, November. Note also the temperature phase-lag, which increases with depth, so that during periods of highest surface temperature, from June through August, the ground temperature at 5m is at a minimum and vice versa for periods of low surface temperature.

An important factor in controlling the energy required to heat or cool a building is the "thermal mass" involved. This is simply the heat capacity which is the amount of energy required to raise the temperature by 1°C. A building having a large thermal mass within the insulation can therefore store a large amount of energy, so that during the day solar energy entering the south windows can be stored within the structure's thermal mass and the temperature will rise slightly and slowly throughout the day. During the night there will be a net heat loss. Heat will transfer from the thermal mass to the inside air so that the temperature will fall slowly through the night. On the other hand a building with a low thermal mass cannot store much energy per degree rise and therefore the incoming solar energy will quickly heat that section of the building to an uncomfortably high temperature, possibly leading to a local cooling requirement. Since little energy is stored, the temperature will fall rapidly during the night and heating will be required.

The thermal mass of soil surrounding a building acts in a similar way and as shown in the previous figures will reduce the temperature variation felt by the exterior of the structure. In addition to these energy benefits the earth protects the building from expansion and contraction and especially from freeze/thaw damage. It is worth emphasizing that in the case of a power failure during extremely cold weather the temperature within an earth protected building will not fall below freezing so that no damage will be done and the building will still be habitable even over long periods. As we know surface buildings in these conditions rapidly drop in temperature, pipes freeze causing great damage and the building is not habitable. Earth protection therefore reduces the need for an external energy source from a matter of survival to one of comfort control. This aspect is particularly noteworthy for remote buildings such as farm buildings, ranger houses and holiday cabins.

In the following discussion the thermal characteristics of various roof, wall and floor configurations are discussed individually and in combination, since each

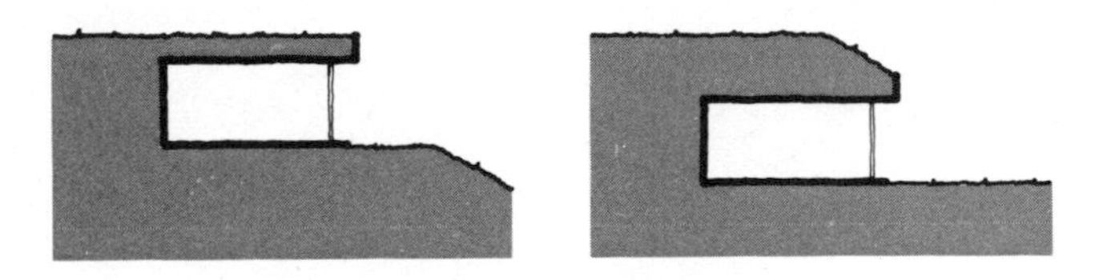

3-3 effect of structure depth

element of an earth sheltered structure interacts with the other components to a greater degree than for conventional surface structures. As an example, varying the amount of earth cover on the roof has the overall effect of changing the depth of the entire structure. As a result not only are the roof's thermal characteristics changed but also those of walls and floor as they are effectively moved into a new thermal environment.

Evaluating the significance of all related factors,and hence, determining the most desirable configuration for a design must be based upon individual consideration of site conditions, local weather patterns, building orientation, intended use, and economics. Therefore, it must be stressed that in the accompanying comparisons, the data given has been generated with respect to design conditions deliberately conceived for each case in order to illustrate the contribution of the particular aspect under consideration. As such, the values given should not be interpreted as design data for any structure in this region but, instead, should be considered as an indication of the general performance of earth sheltered housing compared to an equivalently sized surface structure.

The data developed in this section of the report on the behavior of earth sheltered roofs, walls and floors is used in the energy analysis section in which the total energy requirements of fully earth sheltered, partially earth sheltered and above ground construction are compared.

roof

Due to the reduced energy losses through the walls and floor of an earth sheltered house, the winter heat loss through the roof of such a structure in Minnesota may typically exceed 50% of the total heat losses through the building envelope. As a result, the design of the roof structure warrants particular care and attention.

A primary concern is the determination of the optimal depth of soil, if any, for such a structure. Since the thermal conductivity of soil is approximately 25 times greater than modern insulating materials, its conduction properties are not a significant factor in roof design. In evaluating the heat transfer, it is necessary to examine two other important properties of the total roof system which can greatly influence the overall performance. These are the heat capacity, or thermal mass, of the roof structure and the surface boundary conditions on top of the roof.

In analyzing the heat flux through the roof, various roof cross-sections were examined by means of a transient, one-dimensional, finite-difference computer program (see Appendix A). The inside air temperature was allowed to vary seasonally and ranged from a low of 20°C (68°F) during the winter months to a high of 25.6°C (78°F) at summer's peak. The outside air temperature varied according to a sinusoidal approximation of the average Minneapolis temperatures between the years 1940 to 1970. No solar gain was included at the outer surface as this surface was taken to be covered by grass. The justification for this assumption is discussed later in this section.

A comparison was made between the two alternative roof designs shown in Fig. 3-4 with similar R-values. The first consisted of 3.0m (9.8 ft) of soil supported by a 31 cm (12 in) precast concrete section. The R-value of this design is 4.32 sq m-K/W (24.51 hr-sq ft-°F/BTU). The second was 46 cm (18 in) of soil over 10 cm (3.9 in) of polystyrene insulation supported by a 20 cm (8 in) precast concrete section. In this case the R-value is 4.35 sq m-K/W (24.68 hr-sq ft-°F/BTU). Calculating the energy flux on a daily basis for the entire year reveals that, for the seven month heating season from October through April, the 3.0m of soil offers a 2.4% reduction in heat losses. From these and further studies, it can be shown that in order to compete effectively with standard insulating materials, soil depths in excess of 2.75m (9 ft) would be required on the roof. However, it should be noted that the increased depth of the building would also reduce heat losses through the walls and floor.

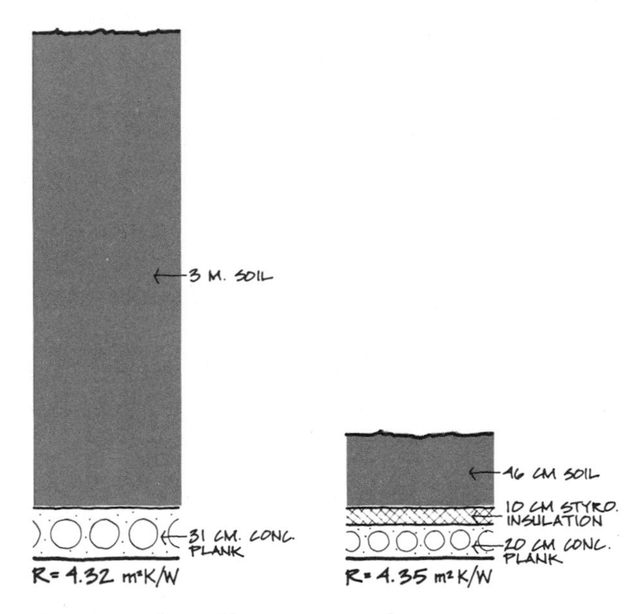

3-4 roof section comparison

While the above procedure serves to predict the energy losses over a period of steady rise and fall of temperature as given by the sinusoidal mean, daily temperatures vary in random intervals of days above and below this mean. The high mass roof structure provided by earth cover damps out or appreciably reduces these temperature fluctuations. As an example a sample January day was selected in which the outside air temperature varied from an early morning low of -15°C (5°F) to an afternoon high of -9.4°C (15°F). On the fifteenth day of the month, a passing cold front was taken to cause a drop of 5.6°C (10°F) for all daily temperatures. Temperatures then remained in this range of -20.6° to -15.0°C (-5 to 5°F) for five days when they returned once again to normal on the twentieth day of the month. The time responses to these conditions of two roof structures with identical R-values of 4.35 sq m-K/W (24.68 hr-sq ft-°F/BTU) are shown in Fig. 3-6.

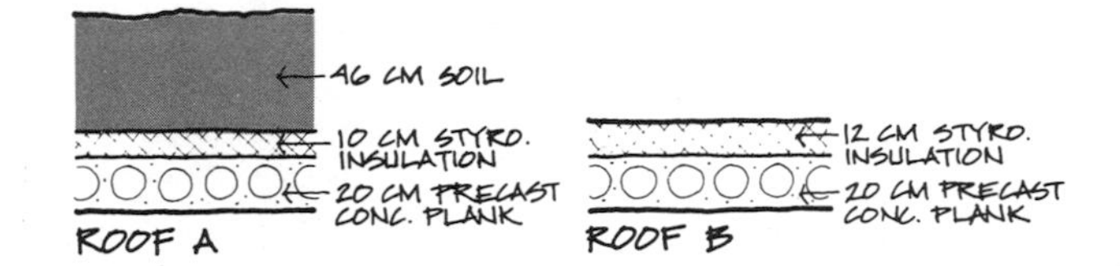

3-5 roof section comparison

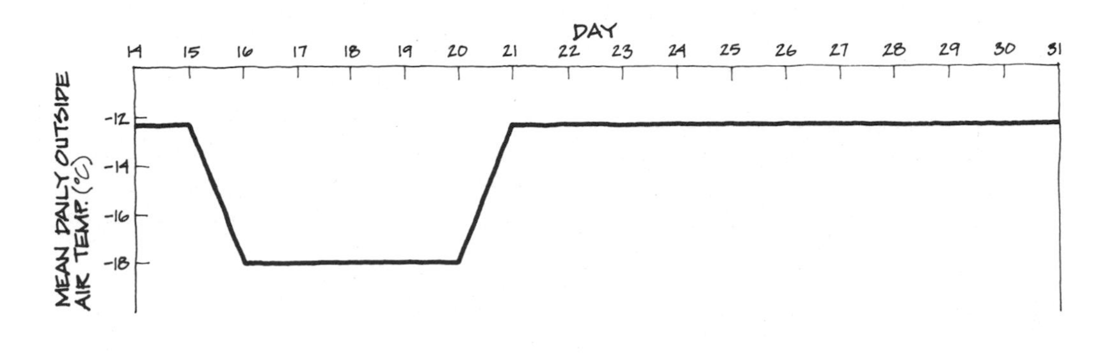

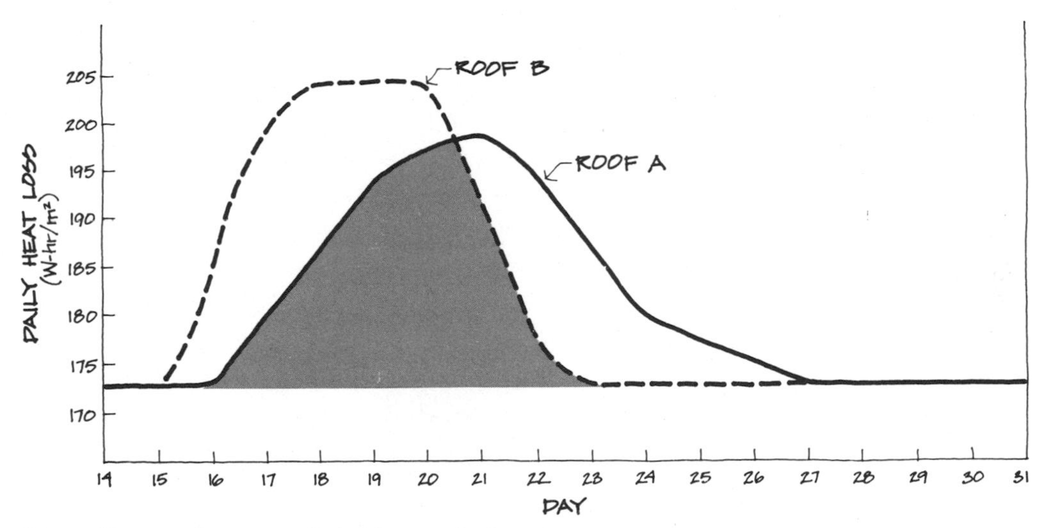

3-6 thermal mass effect in roof structures

Roof A consists of 46 cm (18 in) of earth cover placed over 10 cm (4 in) of high density polystyrene while roof B is a low mass structure insulated by 12 cm (4.6 in) of the polystyrene. Both roofs are supported by identical 20 cm (8 in) precast concrete planks. The daily heat loss per unit area is plotted in the graph, hence, the shaded areas beneath each line represent the excess heat loss due to the drop in outside air temperature for each respective roof type. Due to its low thermal mass, roof B responds immediately to the change outside and within two days reaches a new maximum heat loss which is maintained for the remaining three days. When the temperature returns to normal, roof B responds at once and after two days has returned to its normal January operating level. Roof A, however, requires a full day before the ceiling begins to indicate that more severe conditions now exist outside and once it does begin to respond, does so much more slowly than roof B. After five days, when outside temperatures return to normal, the heat loss of roof A is still gradually rising having attained only 77% of the increase of roof B. Roof A requires another full day before responding to the return of normal weather conditions.

As is shown by the figure, despite the much longer total response time, roof A required 8% less total energy than roof B to cope with the severe period. At the same time, roof A exhibited a peak increase in load which was only 85% of that of roof B. Furthermore, roof B's low thermal mass required almost twice as much additional energy (196%) during the five day period in which outside air temperatures were most severe, thereby placing the bulk of its demand during the period when the outside to inside air temperature differential was at its greatest and required the most energy input from the furnace to bring ventilation air up to room temperature. While this latter consideration does not affect the net energy balance for the structure, it does increase the peak energy demand which must be supplied by the heating system as the building must provide additional energy during a cold wave to warm incoming fresh air independently of the performance of the roof. Thus, it is demonstrated that while two structures may be thermally equivalent under steady state conditions, short term transient fluctuations will cause the high mass structure to exhibit greater stability during operation as well as demonstrate a potential for net energy savings over the low mass structure. With increased earth cover this tendency becomes more pronounced and is a primary factor in the very moderate and stable energy losses in the walls and floor of an earth sheltered building, as will be discussed later on.

A final consideration which comes to light in determining the thermal value of earth cover on a roof is that of the boundary condition which exists at the upper surface of the roof. Vegetation on this surface can contribute to the thermal efficiency in a number of ways, such as shading effects, improved insulation due to air trapped in the foliage, and most important the elimination of solar heat gain to the roof during the summer months. This is mainly due to transpiration which is the cooling of grass by release of moisture. In measurements made of the ground temperatures beneath paved and grass covered patches, Kusuda (see bibliography) observed that, due to solar radiant heating, daily high temperatures during the summer can exceed 60°C (140°F) on an asphalt surface even though the air temperature is no more than 32°C (90°F). In contrast, the high temperature recorded for a grass covered surface under the same conditions attained only 40°C (104°F). An appreciation of this difference can easily be gained by testing the temperature of a sidewalk surface and that of an adjacent grassy area on a clear summer afternoon. Examination of the average surface temperatures to determine the net effect of radiant input shows that during the summer months a blacktop covered surface becomes 8.3°C (15°F) warmer than the average air temperature while grass cover is consistently below ambient conditions by 0.6 to 3.9°C (1 to 7°F) depending upon the length of grass (see Fig. 3-7). During the winter there is no appreciable difference between blacktop and sod surfaces, which is in keeping with expectations since the incident solar radiation on a horizontal surface is negligibly small during those months. The actual temperatures recorded above an earth sheltered structure would be somewhat warmer than those indicated in the figure as the presence of the building would serve as a moderate heat source.

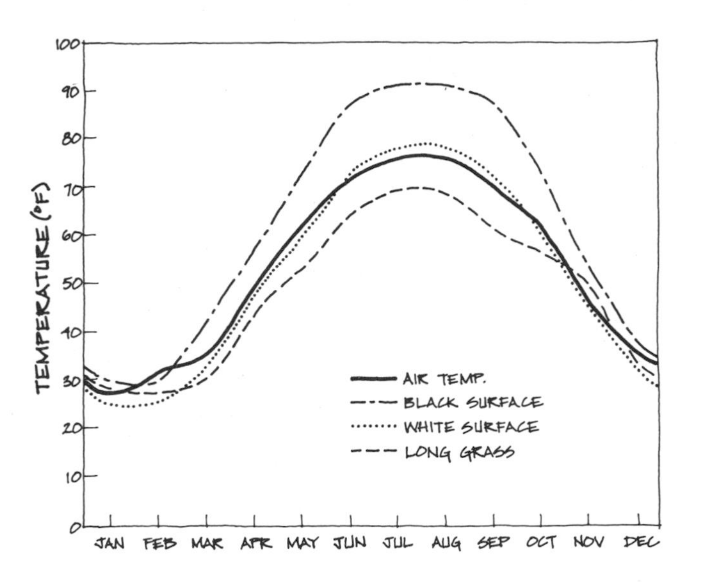

3-7 monthly average surface temperatures

walls

Critical to the inherent advantages of earth sheltered construction is the selection and placement of doors, window openings and wall insulation. During the winter months, a north facing window will lose an average of twenty times the amount of heat as an equivalently sized section of wall on a subgrade dwelling. This disparity can reach a factor of 35 during the severest weather. Likewise, even with the benefit of double glazing and insulated drapery, an east or west facing window represents ten times the heat loss of the same area of earth sheltered wall. A secondary effect of windows or doors is that the opening in the wall diminishes the effectiveness of the earth cover in the immediate area by bringing outside surface conditions up to the wall. During the summer months, even a well designed window represents an undesirable heat gain while the surrounding earth aids the building energy balance by cooling the structure.

South facing windows properly designed to be directly exposed to the sun during winter and shaded in the summer, can actually provide a positive heating effect during the winter months when they are covered by insulated shutters during the night. These gains are roughly equal to the losses per unit area for a conventional above grade wall and can be as much as eight times the wall losses of an earth sheltered section. Therefore prudent design calls for minimizing the window area to the north, east, and west with the majority of windows on the south wall. Further aspects of window design are discussed in the section on passive solar heat gain.

A second consideration which has previously been a point of some contention regarding earth sheltered housing is the question of how much wall insulation is required and where it should be located. One approach has been to place insulation over the entire wall to reduce all heat loss to the surroundings. While such an arrangement appears plausible, particularly during winter conditions, it should be noted that by so doing the building is thermally isolated from the surrounding soil and is therefore unable to take advantage of the thermal mass of its environment. As the building is heated, it will begin to warm the surrounding layers of soil until the soil has reached a temperature ranging between the building temperature and that which existed in the ground prior to its construction. The effect of this is that once the surrounding soil has reached a new steady condition, the building walls will see a smaller temperature gradient, and hence, have lower heat losses. In addition, there then exists a substantial mass of soil surrounding the building which will resist any change of temperature within the building. Thus, with the coming of warm summer months, the surrounding soil will retard any rise in building temperature. Similarly, with the advent of winter this soil mass, which has been warming slowly during the summer, will in turn serve to moderate any downward trend in building temperature.

Except for deep installations in which there is more than 3.0 m (10 ft) of earth cover above the building, the close proximity of the upper half of the wall to the ground surface results in outside temperature conditions dominating the temperature profile of the adjacent soil. As with the roof of the structure, this area must be insulated since the building's heat loss is greatest in this region. For the case in which there is at least 50 cm (20 in) of soil cover on the roof, the lower half of the wall sees an annual temperature variation in the surrounding soil ranging from approximately 6° to 14°C (43 to 57°F). As a result of this, the lower half of the wall is provided with a more moderate environment suitable for use as a source of thermal inertia than is the upper section.

In testing this theory, a sample structure was assumed with the roof insulated by 50 cm (20 in) of soil above 10 cm (4 in) of polystyrene insulation. Two cases were studied in which an identical amount of insulation was applied to the wall (see Fig. 3-8). The difference between the two cases was that wall A had 5 cm (2 in) of polystyrene installed over the entire wall, while wall B had 10 cm (4 in) of insulation placed over the top half of the wall leaving the lower portion exposed to the soil in order to capitalize on the soil's thermal stability. Thus, the two test cases represented identical capital investments with the sole difference being how the materials were utilized. The computer analysis showed that three years were required for each building to attain steady operating conditions with its surroundings. At the end of that period, the model which had insulation over the top half of the wall (wall B) enjoyed a 10% improvement in summer cooling at the cost of a 5% increase in winter heat losses. Averaged over the three summer months cooling season of June through August and the seven month heating season of October through April, this represents a seasonal savings of 100 kW-hr during the summer at the expense of 107 kW-hr during the winter for a 140 sq m (1500 sq ft) home. While there appears to be no net energy gain at first glance, it is significant to note that in wall B the advantage has been shifted towards the warm summer months. The importance of this fact will be clearly delineated in the following discussion of overall building performance where it will be demonstrated that summer cooling is a critical factor in such a structure.

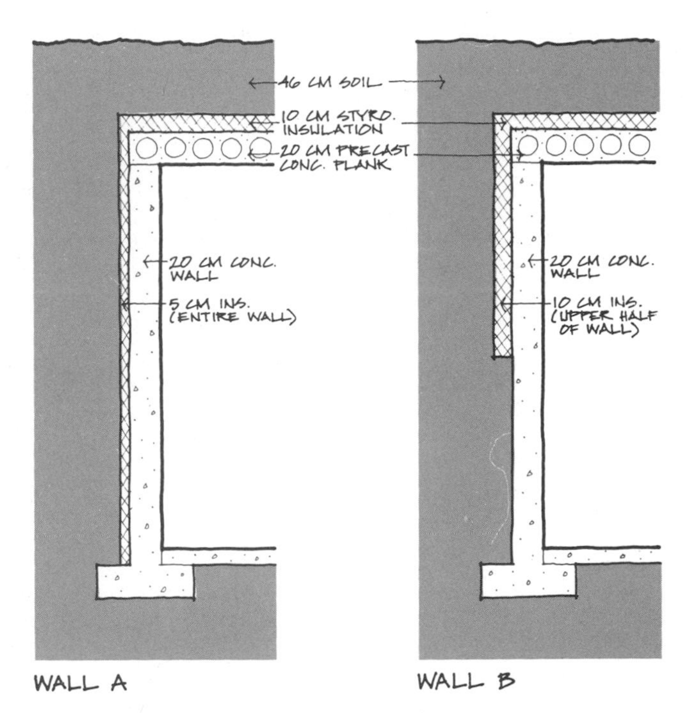

3-8 wall insulation comparison

floor

The floor of an earth sheltered structure is situated in an environment where normal temperatures vary annually within the narrow range of 11°C (42°F) to 9°C (48°F). Thus, the floor of such a structure represents a steady, albeit small, heat loss from the building year around. Computer studies indicate that the rate of loss through an uninsulated concrete floor is on the order of 1.2 W/sq m (0.38 BTU/hr-sq ft) which represents less than 14% of the total winter's energy loss through the building envelope. As the lines of heat flow follow path lengths which are on the order of 10 to 20 m (33 to 66 ft) to the ground surface or deep earth, the addition of 2.5 cm (1 in) of polystyrene insulation beneath the floor decreases the amount of heat loss through the floor by less than 5%, which in turn, computes to be less than a 1% decrease in the total building heat loss.

The same amount of insulation added to a roof with an R-value equal to 4.23 sq m-K/W (24 hr-sq ft-°F/BTU) results in a 20% decrease in ceiling losses during the winter, or an 11% improvement in overall building performance. The conclusion is that from an energy standpoint it is far more cost effective to add insulation to the roof structure than to the floor.

This conclusion is reinforced when examining summertime conditions inside the building. As in the case of the exposed lower wall, when the interior temperature begins to rise due to the warmer incoming air, the thermal inertia of the soil below an uninsulated floor responds by drawing heat out of the building at a greater rate in an attempt to maintain a steady temperature. Thus, a three-fold increase in heat loss through the floor can be encountered during the summer to aid in stabilizing the building temperature within comfortable levels. Installing insulation beneath the floor slab significantly reduces the effective mass of the floor system by isolating the concrete structure from the surrounding soil. Thus, one of the most desirable characteristics of earth sheltered structures is greatly inhibited.

While it can be seen that from an energy standpoint underfloor insulation is a superfluous expenditure, the problem of condensation on cool surfaces and the valid question of human comfort must be considered. To alleviate the cold sensation, underfloor insulation has been advocated to allow the floor slab to float more nearly at room temperature and to isolate it from the cooler deep earth temperatures. While this technique would certainly result in a floor which is at room temperature, 23° C (73°F) there still remains a 14°C (26°F) disparity between floor and body temperature. Thus, while somewhat reducing the sensation of coolness in the floor, it would still be necessary to install insulation in the form of carpeting or wood flooring on the upper surface of the floor slab in order to permit an acceptable comfort level for extended periods of time. In light of the preceding discussion, a better arrangement would be to provide the necessary insulation in the form of carpeting or area rugs rather than the additional under-

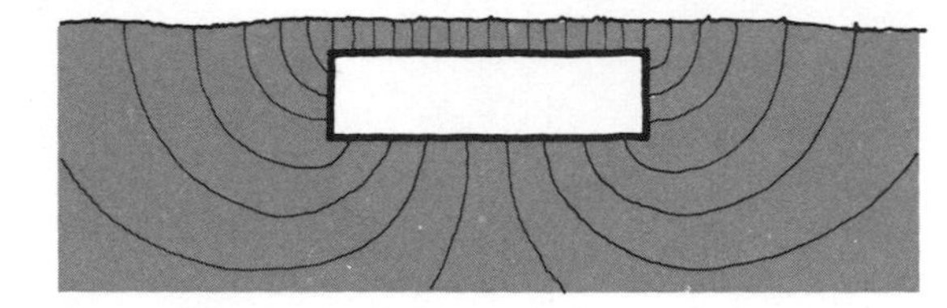

3-9 lines of constant heat flow

floor insulation. Thus, the desired range of comfort would be assured without any appreciable degradation of the positive attributes of the floor structure's thermal behavior.

A final aspect of floor design which must be included in the energy analysis concerns edge losses around the perimeter for walls which have not been earth sheltered. Such a condition exists along the exposed wall of an elevational configuration, as illustrated in Fig. 3-10. In this region, indicated as point A in the figure, the floor slab is essentially equivalent to a slab on grade structure. As is shown in the analysis of a conventional house in the Appendix, this area can create a substantial winter heat loss. Therefore, in order to reduce the influence of the outside weather conditions it is advisable to place insulated footings along exposed boundaries as shown schematically in the figure. A discussion of the effect of various footing and insulation arrangements is given by Algren (see bibliography).

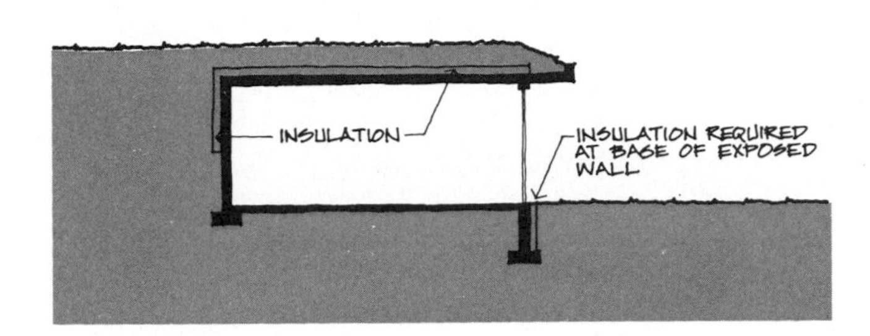

3-10 insulation along exposed wall

overall considerations

From the analysis discussed in the preceding sections, the following criteria are the most favorable for constructing earth sheltered buildings.

The roof of the structure should consist of a grass or foliage cover on a depth of earth sufficient to provide adequate drainage and growth room for the root systems of the plants. Some aspects of roof top planting are discussed in section 2 under landscape concerns. In all cases, however, it is advantageous to work with the maximum depth possible in order to derive the most beneift from the thermal mass of the roof. The limiting factor here is the physical structure required to support such a roof. As the weight of the soil increases, the cost of such a structure quickly escalates beyond the benefits accrued from the additional mass. This can result in the interior spaces being subdivided into smaller areas due to the shorter spans required for high loading factors. Therefore, it is usually preferable to increase the R-value of the roof by means of adding insulation to the structure rather than soil once the load limit of the lighter supporting structure has been reached.

The question of how much roof insulation is desirable was examined by comparing the value of the energy saved to the extra cost of insulation for two roofs (see Fig. 3-11). The first roof consisted of 46 cm (18 in) of soil, 10 cm (4 in) of polystyrene insulation and a 20 cm (8 in) precast concrete supporting structure. The second roof consisted of the same structure with an additional 5 cm (2 in) of polystyrene added to the layer of insulation. The former structure has an R-value of 4.35 sq m-K/W (24.68 hr-sq ft-°F/BTU) which was chosen as a minimum acceptable value while the additional insulation in the second case increased the R-value to 6.01 sq m-K/W (34.10 hr-sq ft-°F/BTU). Based upon the Minneapolis weather data, the additional 5 cm of insulation accrued a savings of 6.67 kW-hr/sq m during the winter and 0.29 kW-hr/sq m during the summer giving a total savings of 6.96 kW-hr/sq m. Using a cost of $0.03/kW-hr for electricity, this equals a savings of $0.21/sq m per year while a projected cost of $0.10/kW-hr for electricity gives a total benefit of $0.69/sq m per year for the added insulation. If the insulation sells for $5.38/sq m per 5 cm of thickness ($0.25/board foot), then the payback period for the extra insulation will be 26 years at the rate of $0.03/kW-hr and 8 years at the rate of $0.10/kW-hr (without any interest charged to the capital investment).

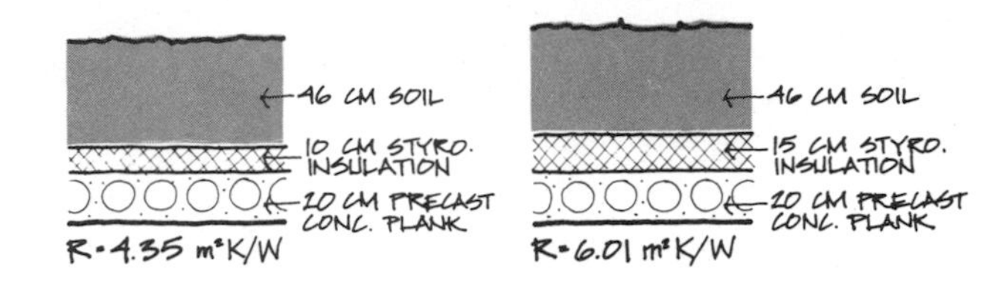

3-11 roof section comparison

The upper portions of the wall should be insulated to a depth of 2.15 m (7 ft) similar to the roof shown in the adjacent illustration. Below this level, the insulation may be eliminated or tapered in successively thinner layers depending upon an engineer's assessment of the building's particular needs. Refer to the section on dehumidification and insulation for a discussion of the possible implications of wall insulation with respect to condensation on the wall surfaces.

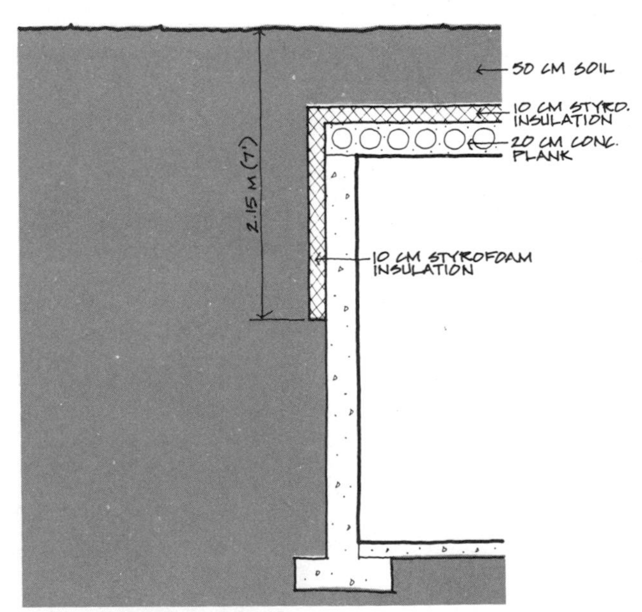

3-12 wall insulation

In lieu of insulating the upper wall surface of the building, there are two alternative configurations which are worthy of note. The first, shown here in Fig. 3-13 extends the roof insulation past the wall before dropping a vertical section of insulation parallel to the wall with soil backfilled between it and the wall surface. This scheme would require an additional amount of insulation proportional to the distance the roof insulation extends beyond the building perimeter The philosophy behind such a modification is to create an insulated cap which fits over the building, and by enclosing soil between the wall and the insulation, inexpensively increase the thermal mass of the building in this critical area. Working with a separation distance of 20 cm (8 in) from the wall, it was found that this arrangement offers a 109 kW-hr savings during the winter and a 100 kW-hr increase in cooling capacity during the summer over the configuration shown in

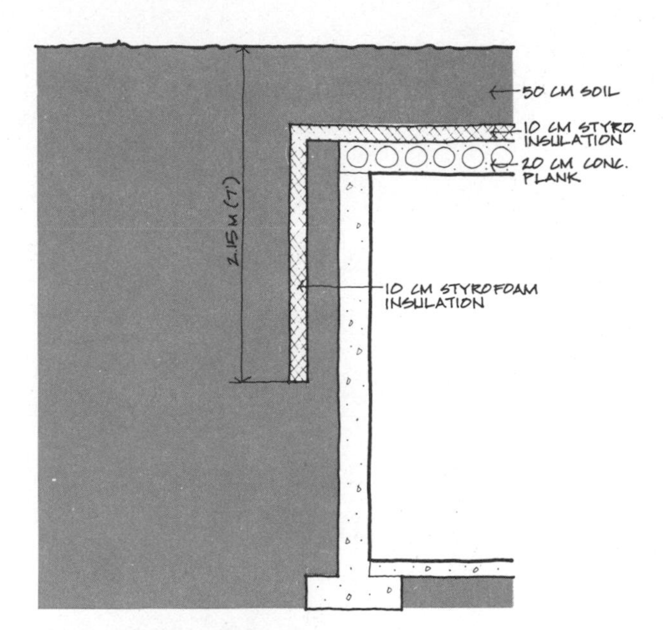

3-13 wall insulation

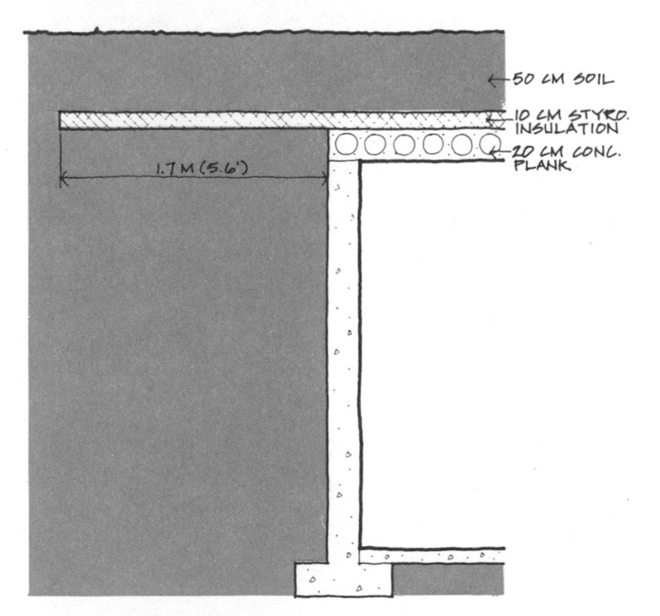

3-14 wall insulation

Fig. 3-12. These figures are taken for a 10 m by 14 m (33 ft x 46 ft) home with 2.5 m (8.2 ft) walls and does not include the effect of window or door openings. The savings of $6.27 to $20.90 per year (based on $0.03 to $0.10/kW-hr), when compared with the estimated additional cost of $133.42 for the added insulation shows that such schemes may become quite competitive as utility rates continue to rise in the future. Of course, if the price of insulation increases proportionately beyond the assumed cost of $0.25/board foot at the same time, then these conclusions must be reassessed. A possible difficulty with this arrangement is the requirement that the insulation must seal well and maintain its structural integrity for the life of the building or else thermal "short circuits" may develop which will render the vertical section functionless and result in a condition which is worse than the wall mounted configuration.

A second alternative insulating scheme is illustrated by Fig. 3-14. Here, the roof insulation has been extended directly past the building a distance of 1.7 m (5.6 ft) for this test case. The intent is once again to effectively increase the thermal mass around the wall without appreciably increasing the cost of insulation. An additional benefit is the provision of a watershed which will prevent water from accumulating around the building walls. In sites where soil conditions permit, this second feature may provide a marked savings in construction costs due to less stringent waterproofing requirements. For the 140 sq m (1500 sq ft) test plan, this configuration demonstrated a decrease of winter heat losses of 54 kW-hr while the summer cooling through the walls increased by 233 kW-hr for the season. Thus energy savings of $8.60 to $28.70 per year are possible at the cost of $215.00 for insulation at the corners. As with the configuration in Fig. 3-13, the necessity of maintaining the structural integrity of the insulation may prove a serious handicap to this plan. In addition, both the arrangements in Figs. 3-13 and 3-14 may be seriously compromised not only in terms of energy savings but also in terms of a more complicated construction by the location of windows in the walls. Therefore, these techniques are best suited for structures which concentrate the windows and doors along a single wall.

At this point, it is worthwhile to discuss briefly the physical reasons for the variations in heat transfer characteristics between the configurations shown in Figs. 3-12, 3-13, and 3-14. In Fig. 3-13 the thermal resistance of the upper wall is increased slightly by the insertion of a soil layer between the wall surface and the insulation. More importantly, as the largest surface area of this region is either facing the building or the insulation, the soil contained within the enclosure will tend to approach the building operating temperature and then maintain that level due to its thermal mass. The area is shielded from outside surface changes by the insulation, and at the lower end, where the soil is not insulated, the variations in surrounding temperatures are moderated by the depth, the presence of the building and the relatively small surface area of the exposed face. Thus, one would anticipate a general improvement in both summer and winter operating conditions. Fig. 3-14 also displays a means of improving the effective thermal mass of the wall and increasing the R-value by forcing the paths of heat flux to extend over much greater distances for the upper wall sections. A principal feature of this last configuration, however, is that the wall and its surrounding layers of soil are more directly open to the influence of the cooler deep earth regions. Thus, an improvement in the winter heat loss conditions is expected, but the greatest contribution from a simple extension of the roof insulation is realized in the form of increased summertime cooling due to more effective use of the thermal mass of the surroundings.

A final variant of single story construction is the effect of berming the earth around the outside walls as opposed to a fully recessed configuration as shown in Fig. 3-15. This building mode is particularly suitable in areas where deep excavation is either economically or geologically undesirable or in locations where the general texture of the landscape favors a gentle slope around the building. For the trial case of 140 sq m (1500 sq ft) home with a south facing exposed wall and three earth sheltered walls, the bermed structure showed a 5% increase in heat loss both during the winter and during the summer when compared to a fully-recessed structure. Thus, when considering the total gains accrued by means of earth sheltered construction and allowing for slight error due to the simplifying assumptions in the analysis, it can be seen that the totally submerged subgrade structure enjoys marginal advantage over the bermed structure, and certainly not by a sufficient degree to warrant major modifications in the building.

FULLY RECESSED STRUCTURE

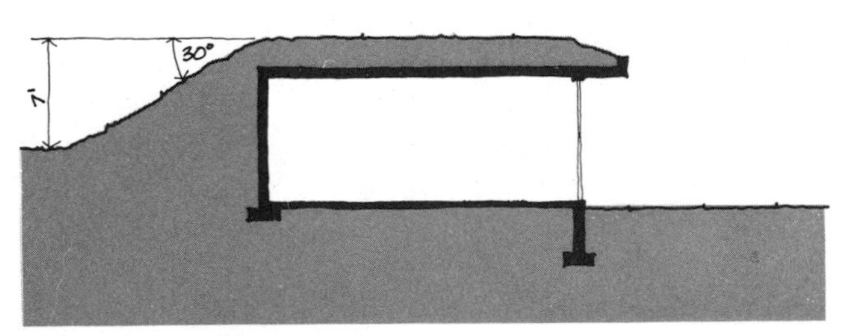

BERMED STRUCTURE

3-15 effect of berming

The items discussed in this section of the report represent a variety of concerns that are relevant to the energy use and mechanical design of earth sheltered housing. The first three topics discussed, sources of heating and cooling and their method of delivery, are general in nature and can apply to any type of structure. The next two concerns are ventilation and humidity control which are particularly important to earth sheltered design. The final section deals with HVAC controls when applied to this type of structure.

heating sources

Although the following items are quite different in nature, they all represent basic sources of heat available in a residential structure. In the case of passive solar energy, and internal heat sources, information is developed which is used in the total energy analysis of earth sheltered houses at the end of this section.

conventional systems

The majority of homes and heating systems use fossil fuels. Fossil fuels include oil, natural gas, L.P. gas and coal. Electricity, which is produced primarily from fossil fuels is also used commonly for home heating. The reasons for the popularity of these fuels include low cost, cleanliness, automatic operation and low investment required for equipment. All of these conventional heating systems can be adapted to earth sheltered homes. As fuel costs increase many alternative heat sources also become feasible for earth sheltered homes because of the reduced energy requirement. Descriptions of some alternative heat sources follow.

heat pumps

As the popular fossil fuels become unavailable for new homes, more people must rely on electricity for heating in which case greater efficiencies and vast savings can be realized through the use of electric powered heat pumps. Such units may be either air coupled or liquid coupled. Because of the low heating requirements of an earth sheltered structure, the capacity and cost of this equipment could be significantly reduced, but the physical size would not decrease noticeably. Presently, the COP (coefficient of performance—a measure of efficiency), for such equipment varies between 2.5 to 4.0 (i.e. for each unit of electrical energy input, 2.5 to 4.0 units of heat energy are transferred from an energy source to the interior home space). On the other hand, the COP of electrical resistance or radiant energy is 1.00. Hence, the heat pump is 2.5 to 4.0 times more efficient than other forms of electrical heating. In addition, the heat pump automatically includes air conditioning and can be nicely coupled with various solar heating systems.

wood

Wood is becoming an increasingly popular fuel for supplementary heating, and is used in some instances as a primary fuel. This new popularity has created a demand for improved wood burning equipment that is being met by designs with increased efficiency and convenience. Wood fired furnaces and boilers are available and can provide controlled heat with minimal refueling. Some models even provide for automatic gas or oil backup so that heating may continue uninterrupted as the wood burns itself out.

Although wood is considered inconvenient as a primary fuel for a home with a large heating requirement, the low heating requirement of an earth sheltered home makes the use of wood feasible, especially if a method is provided for storage of the excess heat. A wood fired boiler coupled to a storage tank provides a heating system which can heat a home with only periodic fueling. If this system is used the stored heat must be delivered to the space to be heated through either radiation units or a forced air coil. This system can provide increased efficiency for the wood boiler as well as yielding a more even heat distribution to the space.

internal heat gain

Another source of heat available in a home comes from within the structure in the form of internal loads. This heat occurs automatically in all residences and reflects the size and lifestyle of the occupants. These loads are listed in Fig. 3-16. The values in this table for daily internal heat gain (27.8 kW-hr/day in winter and 18.4 kW-hr/day in summer) are used in the total projected energy use for various test structures in the energy analysis section. Please note the assumptions when using these figures.

dryer (vented inside)	5000w @ 1 hr/day	5.00 kW-hr
refrigerator (average)	300 w @ 24 hr/day	7.20
television	200 w @ 3 hr/day	.60
water heater	150 w @ 24 hr/day	3.60
cooking	2000 w @ 2 hr/day	4.00
lights (average)	3 w/sq m x 280 sq m @ 4 hr/day	3.40
people	1 person x 60 w x 8 hr/day	
	x 75 w x 6 hr/day	.90
	1 person x 60 w x 8 hr/day	
	x 80 w x 12 hr/day	1.40
	2 people x 45 w x 8 hr/day	
	x 60 w x 8 hr/day	1.70
		27.8 kW-hr/day

3-16 internal heat gains

notes: 1. 27.8 kW-hr/day is the winter condition.
18.4 kW-hr/day is the summer condition–dryer heat is vented outside, less energy is used for lights, cooking, tv, etc., because of outdoor activities.
2. Values are obtained from section 2-99 to 2-102 of the Handbook of Air Conditioning, Heating and Ventilating Strock and Koral, Second Edition.
3. Values are very dependent upon life style, family size and geographic location. Reference: Smith N., **Energy Management and Appliance Efficiency in Residences,** Proceedings of the International Conference On Energy Use Management, Vol. 1, Oct. 1977.
4. With the advent of newer energy efficient appliances, the values listed may decrease somewhat.
5. All appliances are assumed to be electric.

active solar systems

A great amount of information on solar heating systems has recently become available. There are a few implications of solar heating to earth sheltered residences which should be considered. The major drawback to solar systems in surface residences is the large capital investment required. In an earth sheltered residence the reduced heating requirement allows for a correspondingly lower investment in collecting equipment for solar heating. In addition, the storage requirement for a solar system is reduced in proportion to the heating requirement and further because of the large thermal mass of the house itself. This reduction in storage volume also results in a reduction in the floor area required for installation of the storage media. A comparison of storage materials is given in Fig. 3-18. Water and rock are the only media currently used widely. Wood and oil are included to give a comparison to storage of fuels which are burned.

material	volume	usable stored energy	storage density
water 70°F Δ T	128 cu ft	560,000 BTU	4,375 BTU/cu ft
rock 70°F Δ T	128 cu ft	270,000 BTU	2,100 BTU/cu ft
hardwood	1 cord	10,460,000 BTU	81,718 BTU/cu ft
softwood	1 cord	5,490,000 BTU	42,890 BTU/cu ft
oil	128 cu ft	100,000,000 BTU	785,000 BTU/cu ft

conversion efficiency of wood = 50%
1 cord = 4′ x 4′ x 8 = 128 cu ft, moisture content = 25%

3-18 energy storage capacity of materials

Water storage is the most flexible from a mechanical systems point of view, however, its average installed cost is very dependent upon the size and type of tankage used. Rock doesn't leak and may be used directly with forced air heating systems. Rock storage may or may not be more expensive than water depending upon the distance it must be hauled to a particular site and may develop long term odor and bacteria problems due to the dark, moist and warm rock surface area.

The use of active solar collection to heat domestic hot water appears to be a very viable alternative. Presently manufacturers are offering such units, with 1977 prices ranging from $800 to $2000, depending upon size, performance and options. The payback time period for the typical unit when based on electrical water heating rates, averages 4 to 6 years. The heating of domestic hot water accounts for approximately 15% to 20% of the total energy consumed in a typical 4 person surface residence. In an earth sheltered residence with its low heating requirement the heating of domestic water can become the largest heating requirement in the house. Information on manufacturers of solar domestic hot water systems can be found in the reference section.

passive solar heating

Direct solar radiation through the windows of a house is an important source of heat. Although some radiation is available in east and west facing windows, south facing windows provide the maximum amount of passive solar heat. In order to illustrate that properly designed south windows can provide net heat input to a house, calculations were made based on the following assumptions. Wall construction Y as described in Appendix B will be used. The wall is constructed of 2 x 6 framing members on 24 in centers with 5-1/2 in fiberglass batt insulation giving a U-value of 0.182 W/sq m-°C (0.032 BTU/hr-sq ft-°F). The windows are 1.2 m high with the top of each window located 0.5 m below the ceiling, and with an 0.9 m exterior overhang along the entire south wall at 2.5 m above the floor line. In the calculations done for heat loss through the south windows it was assumed that tight fitting drapes with good insulating characteristics were open during daylight hours and closed during hours of darkness, changing the U-value of the window from a daytime value of 3.349 W/sq m-°C (0.59 BTU/hr-sq ft-°F) to a nighttime value of 2.555 W/sq m-°C (0.45 BTU/hr-sq ft-°F).

It should be noted that on a south wall a double glazed window is preferred over a triple glazed window assuming insulated drapes are used. The double glazed window allows a solar energy transmittance value of 72% whereas the triple glazed window only allows 61% transmittance. It should also be noted that in either case insulated drapes tend to prevent the warmth of the room from reaching the inner window pane. This may allow the window temperature to fall below the dew point of the interior air resulting in condensation on the windows. Listed in Fig. 3-19 are total heat gain/loss quantities for winter months through the combined south wall and windows for window areas varying from 15% to 40% of the wall area.

Examination of the values in Fig. 3-19 shows that for winter months, except December, as the amount of window area in the wall increases the heat loss through the entire wall decreases. Table C-1 in Appendix C gives values of heat gain/loss through south facing windows and shows that for all months except December the net energy flow through the window is positive. Increasing the window area (for all months except December) decreases the net energy loss through the wall for winter months until, at 25% window area, the net energy flow for the seven winter months is positive. Totals are also given for November through April to show that even if the large October gains are ignored the net energy flow for winter is positive for window areas of 40% and greater.

	window percentage of total wall area					
	15	20	25	30	35	40
january	−155	−137	−120	−103	− 86	− 69
february	− 54	− 15	+ 25	+ 64	+103	+142
march	− 47	− 14	+ 20	+ 52	+ 86	+119
april	− 20	0	+ 21	+ 41	+ 62	+ 82
october	+232	+332	+431	+531	+630	+730
november	− 37	− 7	+ 23	+ 53	+ 83	+113
december	−184	−185	−186	−186	−187	−187
oct.-apr. totals	−265	− 26	+214	+452	+691	+930
nov.-apr. totals	−497	−358	−217	− 79	+ 61	+201

note: positive values indicate heat flow into the building, negative values out.

3-19 heat gain/loss through south wall/window combination

The information in this table shows that when properly designed and controlled, windows on the south side of the building can provide a net energy gain to the building in winter. There is no apparent limit on the amount of south window to install, although careful examination should be given to each individual building to be sure that windows can gather heat and that the heat can be used effectively in the building. The actual energy gains indicated in the table are based on the assumption that all of the radiation which passes through the windows is retained within the structure. In reality, it may not be possible to absorb and use 100% of the available passive energy. The greatest amount of heat will be absorbed when the sunlight hits surfaces which are dark and objects which are massive.

The figure below illustrates the impact of passive solar heating by comparing the heating requirements of an earth sheltered house on a sunny and a cloudy day in January. The basis for the data is presented in the energy analysis section which follows.

	sunny day kW-hr/day	cloudy day kW-hr/day
net heat loss	−34.68	−67.85
ventilation heating	− 7.35	− 7.35
sub total	−42.03	−75.20
internal gains	+27.80	+27.80
(net loss) total	−14.23	−47.40

3-20 winter heating requirements–subgrade elevational structure.

Note that the amount of heat required to make-up the total net heating requirement of the house on a sunny day is only 14.23 kW-hr. Further analysis will show, however, that there is overheating during the day and underheating at night. In order to maintain an even temperature throughout the day, a thermostatically controlled HVAC system is required.

heat recovery

One source of heat in the home which is often not used is heat recovery from waste water and exhaust air. A significant amount of heat is recoverable from waste water which is normally dumped directly into the sewage system. Heat recovery from exhaust air is less practical than from waste water, but still possible in an earth sheltered home. The following calculations are based on the 140 sq m (1500 sq ft) example house size occupied by a family of four, and are intended to show the amount of heat recoverable. These calculations should not be assumed to apply to all houses without back-up data.

Assuming that a family of four uses about .31 cu m/day (82 Gal/day) of hot water at 54°C (130°F) and that 50% of that heat could be recovered to preheat cold water at 13°C (55°F), then approximately 230 kW-hr of energy per month could be recovered. For purposes of placing a numerical value on the energy recovered a direct heat transfer method has been assumed. It is important to note that for 50% of the energy for heating domestic water to be recovered, a certain investment is required. Careful examination should be given to the cost of installing and maintaining this equipment to insure that a reasonable pay-back is achieved.

The method of recovery of this energy has a significant effect on the recovery efficiency. As noted, direct heat transfer has been assumed at 50%. This figure could be raised to approximately 75% if a liquid coupled heat pump were to remove the energy from the waste water. The total recoverable heat would then become 15 kW-hr/day (5100 BTU/day) a significant portion of the total heating requirement. This approach, however, would only be economically feasible when a heat pump was on site for reasons other than heat recovery. The secondary effects of dumping cool water to a septic system or central sewage system should be studied. Several projects utilizing this approach are presently operating in Minnesota; however, the projects have not operated long enough to yield accurate, extensive data.

If similar efficiencies of heat recovery are assumed for exhaust air, approximately 108 kW-hr of energy per month could be recovered. This is based on 2 air changes per day in the same 140 sq m house with average January indoor and outdoor temperatures. The recovery of heat from exhaust air to heat make-up air is not as simple as recovery from waste water. Problems arise from the fact that air should be exhausted from the kitchen, one or more bathrooms and possibly a general exhaust. Make-up is provided from infiltration, door openings and possibly a general fresh air fan. The transfer of heat from many exhausts to a central, pressurizing fresh air supply is possible but probably not economical. The advantages of this are further reduced when considering that the example was for January and that the energy savings in other months would be less. See the section on ventilation for further information.

cooling sources

The need for cooling in an earth sheltered house is questionable. The test designs presented in the Energy Analysis section which follows indicate cooling loads on the order of 850 kW-hr for an entire cooling season for one and two story sub-grade elevational structures. Even if it is assumed that this entire cooling load occurs in one period of three weeks it represents a cooling load of less than 1/2 ton of refrigeration (1 ton is equivalent to 12000 BTU/hr). Although most people would probably choose to do without cooling equipment in a residence of this type there are several methods, both traditional and non-traditional, to accomplish cooling in an earth sheltered residence.

conventional systems

Conventional direction expansion (DX) refrigeration systems are certainly adaptable to an earth sheltered house. Careful design of the system should be done to insure that it is not greatly oversized. An oversized system will run inefficiently and will result in poor humidity control. These same considerations apply to heat pumps which operate in a manner similar to traditional DX systems. Since an earth sheltered house does not feel the peak summer loads that an above grade house does, the mechanical system can be designed for a smaller capacity. Thus, in addition to lower energy costs, the initial cost of the equipment can be reduced as well.

annual ice storage system

The Ice Air Conditioning System recently developed by Dr. Thomas P. Bligh provides a potential alternative source of cooling. This method is a regional solution and will not function everywhere. The system uses the frigid winter air to freeze a large tank of water into ice. During the summer a glycol based heat transfer solution carries unwanted heat from the house to the tank, thereby melting the block of ice. The energy required to do such is small consisting only of the amount necessary to run a small (approximately 1/5 hp) centrifugal pump and the furnace blower. It should be noted that this system couples nicely with the solar reheat process for precise control of interior relative humidity during the summer months. This system can be used as a primary cooling source where cooling is considered necessary. The major economic consideration is the cost of the storage tank and heat transfer coils. A prototype of this system is presently being tested.

heat loss to the earth

An important benefit of earth sheltered construction is the cooling effect in summer of the earth against the walls and floor. Examination of the heat losses and gains through the walls and floors during the summer months, as given in the Energy Analysis section of this report, will give an indication of the magnitude of this cooling effect. It is the main factor in causing the total cooling load to be so low. This cooling effect could be increased by placing earth against a greater wall area. However, if the earth is placed against the south wall in place of windows, this additional summer cooling would be at the sacrifice of winter passive heating. It is important to consider the annual energy consumption of a residence when determining the amount of earth sheltering.

ventilation during cool periods

As with surface houses, earth sheltered houses can achieve a large amount of cooling with night-time ventilation. In addition, because of the high mass of an earth sheltered house it is possible to cool by ventilation during cold days and retain this cooling through a period of several warm days. It is important to note that in all the calculations done for south windows throughout this study, the drapes were assumed to be open during all daylight hours. In spite of the window overhang a large amount of radiant heat still enters through the windows in summer. Much of this heat can be excluded from the house by closing drapes when the sun is shining in. This will be of great assistance when cooling by night-time ventilation is practical. Refer to the section on Ventilation which follows.

roof spray systems

Vegetation is often present on the roof of earth covered structures for erosion control, reduction of the ground temperature, and aesthetic reasons. Supplying moisture may be necessary to sustain plant growth during dry periods. The use of a properly designed roof water spray system could provide irrigation for the vegetation and create conditions conducive to evaporative cooling. However the effects of such cooling are limited to the soil near the surface. Studies have indicated that the cooling effect of such a system on a building of high mass and good insulation is minimal. Thus, this technique is not likely to provide any energy savings in earth sheltered designs which place earth on the roof.

methods of heating/cooling delivery to building

Several basic systems are typically used for the delivery of heating and cooling to living space. These include the forced air duct system, hot water baseboard system, gravity air system, and radiant system. All these systems have advantages and disadvantages which affects their suitability for an earth sheltered home.

A properly designed forced air system utilizing low air supply temperatures appears to offer the greatest number of advantages, with the fewest disadvantages. It incorporates a great many standard features while allowing several significant additions. The advantages and disadvantages of a forced air system are listed below:

forced air heating advantages

- Filtration is possible with a simple media filter.
- Can distribute controlled outside "make-up ventilation air".
- Can supply heating or cooling.
- Can return air from ceiling space to reduce unwanted indoor air stratification.
- Works well with alternative energy systems such as solar and wood heating.
- Utilizes a "duct only" system outside the mechanical space, and water leaks are not a hazard.
- Can distribute dehumidified or humidified air.
- Air movement may provide desirable background sound level in a quiet, draftless home.
- Provides low energy use with low supply temperature.
- Cost is very competitive when air conditioning or dehumidification is desirable.

forced air heating disadvantages

- Could be overly noisy if not properly designed and installed.
- Probably more expensive on a first cost basis in large homes.
- Can be inefficient if not properly designed or installed.
- Is not as conducive to individual room control as other systems.

options

- Constant fan operation.
- Charcoal filtration for odor control.
- Heat recovery.
- Radiant floor or ceiling distribution system.

Compared to a forced air system, the hydronic (hot water) approach yields a quiet system of moderate first cost when heating alone is considered for a moderate to large size home. In addition, a hydronic system can supply heat to spaces with individual room control; however, it cannot distribute, filter or condition ventilation air. Because of the high operating fluid temperature (approx. 71° to 82°C or 160° to 180°F) the system does not efficiently couple with alternative energy systems such as solar collectors.

The gravity air system, yields a quiet, low cost heating system that can supply humidified air only. In addition, it can work efficiently with properly designed alternative energy systems and it can continue to work during power failures. However, this system generally requires a network of large ducts and cannot filter, cool or dehumidify the supply air.

The most familiar radiant heating systems consist of either an electric grid embedded in the ceiling or floor of a room, or hot water piping embedded in the floor. With the advent of forced air heating systems that utilize low temperature supply air (29° to 35°C or 85° to 95°F) a unique radiant floor and/or ceiling system is possible that increases the comfort and efficiency of the overall heating system by utilizing the supply air as the working fluid of the radiant system. Basically, the floor or ceiling structure is composed of hollow precast, prestressed concrete planks, in which the cores are utilized as the ducts required to distribute the warm supply air. It has been reported after testing in Sweden that the temperature of this type of radiant concrete system averages 2° to 3°C warmer than average room temperature. Because the heat is uniform throughout the entire floor or ceiling area, the room air temperature can be lower than normal while maintaining the same comfort level. In addition, the floors are much more comfortable, and they provide massive internal energy storage that is slightly above room air temperature. It appears that this combined system yields a great number of advantages. Several such systems are presently operating in Minnesota.

ventilation

In a typical above grade house, ventilation is not normally considered a problem since the air infiltration around doors, windows and through walls allows more fresh air to enter than is required. However, tightly constructed houses often eliminate most of the air infiltration to save energy. Fresh air must be provided in some manner. Earth sheltered structures in particular offer the occupant the benefit of minimal unwanted air infiltration; hence, one may control the amount of exhaust and make up air required by ventilation. Ventilation is necessary for the following reasons:

- To supply the proper amounts of oxygen for the health and well being of the occupants.
- To supply the proper amounts of oxygen necessary for combustion if open flame furnace, fireplaces, etc. are on the premises.
- To dilute or eliminate excessive moisture in the air during the summer.
- To dilute or eliminate odors generated in the bathroom and kitchen.
- To dilute or eliminate the heat produced by internal sources during the summer.

In order to make energy use projections for earth sheltered housing, it is necessary to establish a reasonable level of ventilation air which must be provided. The level of ventilation will be determined for a 10 m x 14 m x 2.5 m test house with a total volume of 350 cu m (12357 cu ft). It is assumed that the house is an all electric residence (no open flames) and is occupied by four people.

Of primary concern is the respiratory requirement for the occupants of a house. Generally man needs 20% oxygen in the air. He can exist with 15% oxygen in the air but combustion will not occur. Death for humans will result with only 5% to 7% oxygen in the air. The adjacent table indicates man's oxygen and air requirements for various activities.

If the four occupants are assumed to engage in activities of the .014 cu m/min level for 16 hours per day and of the .006 cu m/min level for 8 hours per day the minimum ventilation level for the house would be 65.3 cu m/day. This requires a complete air change to the house only once every 5.5 days. The 1976 Minnesota Energy Code states: "The quantity of outdoor air introduced into spaces conditioned for comfort for the purpose of meeting normal respiratory and odor control needs shall be no greater than .14 cu m/min (5 cu ft/min) per person." With four occupants and .14 cu m/min (5 cu ft/min) per occupant the ventilation

activity	oxygen consumed cu m/min	air required cu m/min
sleeping	.00024	.006
sitting	.00030	.007
standing	.00036	.008
walking–2 mph	.00065	.014
walking–4 mph	.0012	.026
jogging	.0020	.043
max. exertion	.003 to .004	.065 to .100

3-22 oxygen and air requirements

rate for a house is 815 cu m per day. This results in about 2.3 air changes per day.

Although data is unavailable for determining correct ventilation levels for odor and humidity control, some observations are of value. Data available for infiltration through window cracks and door openings indicates a ventilation level in a relatively tight house of approximately 2 air changes per day. Observations of actual houses which fit these conditions show this to be about a minimal level for elimination of lingering odors, especially pungent cooking odors. The 2 air changes per day ventilation rate is just below the code maximum of 2.33. For purposes of further discussion it will be assumed that the ventilation requirement of a house is 2 air changes per day. This rate should not, however, be taken as the requirement for all earth sheltered houses. Until sufficient experience is gained in the ventilation of these houses, each should be analyzed before construction and provisions should be made for increasing or decreasing ventilation as necessary.

Where open flames, including fireplaces, are present in well sealed homes, increased ventilation must be provided. For purposes of energy conservation, combustion air should be ducted to furnaces or fireplaces. As an alternate solution, delivery of heated make-up air to the proximity of the fireplace may be considered. Ventilation may also be required for the removal of excess internal heat.

Once a ventilation rate for an earth sheltered house has been established, the next step is to determine how that ventilation is to be achieved. It has been stated that a forced air heating system appears to be the most advantageous for this type of residence. A major reason for that decision was the ease of providing make-up air with this system. If a system other than foced air is used a small forced air make-up system should be provided. Including an electric heating coil in such a system would be most convenient. For the remainder of this discussion it will be assumed that a forced air heating system is to be used.

There are several methods available to introduce fresh air into a house with a forced air heating system. The simplest and probably the most economical is to provide a duct from the outside to the return air plenum. The fan will pull in fresh air automatically. A balancing damper should be installed in the duct to control the proper air volume. Exhaust to match the fresh air supply can be achieved through bathroom and kitchen exhaust fans with a general exhaust fan to make up the estimated difference. Refinements to this system can provide finer control of exhaust and make-up volumes. With the development of inexpensive computerized processors, a control system can be manufactured to monitor bathroom and kitchen exhaust volumes and to control the general exhaust and fresh air intake. It should be noted that no attempt should be made to recirculate bathroom exhaust. Many such odors are ammonia based and are not easily filtered or otherwise controlled; hence, exhausting is essential.

At this point the subject of heat recovery from exhaust air can be re-examined. Assuming that the 700 cu m/day (2 air changes) are exhausted in a controlled manner, and that 50% of the heat can be recovered, a projected yearly savings of $32.25 can be made. This is based on the additional assumptions that the indoor temperature is held at 20°C (68°F), outdoor temperatures are as shown in table B-1 of Appendix B for October through April, and that electricity costs $0.06/kw-hr. If a pay-back period of 7 years is considered then only about $225 can be spent for the equipment to realize this energy recovery. Under these conditions, the monetary savings are not large enough to justify the installation of equipment to return 50% of the exhaust air heat.

If heat recovery becomes economically feasible and is desired, the system shown in the adjacent drawing will provide localized heat recovery on the bathroom and kitchen exhausts. The heat exchanger can be a metal unit commonly called a "z-duct." The two fans should be the same size and should run together. The electric heater can be eliminated if the fresh air is ducted to the return air plenum. The fans can be operated by a simple wall switch which could be limited by a timer.

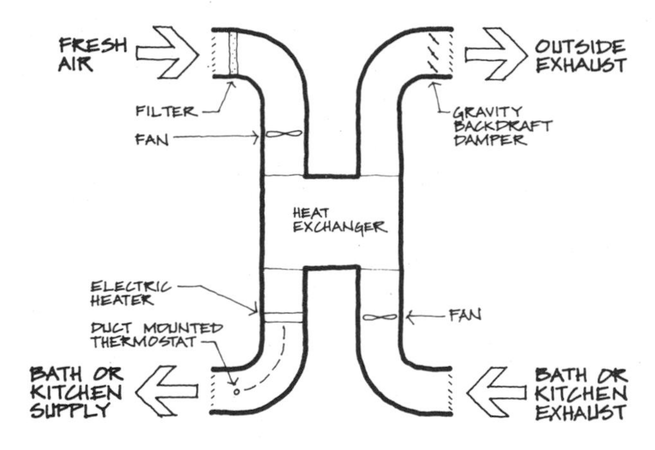

3-23 heat recovery diagram

A final important ventilation consideration is cooling during the summer. In general those ventilation systems already mentioned are best operated at night in summer and during the day in winter. In addition, it may be very advantageous to have a system for providing high rates of ventilation with outside air during the night in summer. By design, an earth sheltered house has a large mass. This large thermal mass will allow substantial amounts of heat to enter the house in summer without overheating the spaces. This effect can keep the house comfortable if the heat is allowed to escape outside at night except during periods of extremely hot weather. To accomplish this some type of forced ventilation system will be required in many earth sheltered designs because cross ventilation is not possible. A forced air heating system can be provided with ductwork and dampers to introduce 100% outside air through the furnace at night with exhaust through open windows. This may require some automatic controls. Where a forced air system is not installed an exhaust fan can be arranged to expel air, causing make-up air to flow in through open windows.

humidity control

Humidity is water vapor that is intimately mixed with air. Humidity control includes both humidification which is the addition of water vapor to increase the humidity level in the air, and dehumidification which is the removal of water vapor to decrease the humidity level in the air. Relative humidity is a term used to compare the amount of actual water vapor that is mixed with air at a given temperature with the maximum amount of water vapor that could be mixed with the air at that temperature. The control of humidity is very important to the well being of the structure and its occupants. The reasons for this include:

- Rusting of unprotected ferrous metals may occur if the relative humidity exceeds approximately 40%.
- Excessive static electricity build-up occurs on synthetic rugs and furniture if the relative humidity is significantly below 55%.
- Wood furniture tends to "dry-out" and crack with relative humidities significantly below 30%.
- People can feel uncomfortably "hot" during the summer if the relative humidity is high, because evaporative cooling of the body due to sweating is surpressed.
- People feel more comfortable and do not develop cracked or dry skin problems if the winter time relative humidity is not too low.
- If the relative humidity is high and cool surfaces are present, condensation will occur. This often occurs on the inside surface of window glass during the winter. Damage from this occurs when the water eventually runs onto the wooden window sill and wallpaper.

dehumidification

Dehumidification is most generally required during the summer when outdoor levels of humidity are high. Additional humidity inside the house is usually unwanted but, nevertheless, produced by the following sources:

- Bathing and showering.
- Dish and clothes washing.
- Cooking.
- Perspiration and respiration.
- Miscellaneous sources—cleaning, standing water, plants, rain water carried in, groundwater seepage, etc.

The need for dehumidification must be determined on an individual basis since the sense of personal comfort varies. The cooling effect of the earth will produce temperatures comparable to those of a surface house with mechanical air conditioning but without the associated dehumidification. This will cause relative humidities in earth sheltered houses to be generally higher so that a dehumidification system may be necessary.

In addition to dehumidification for comfort there is the possibility for high humidity levels to cause condensation and subsequent damage on cool surfaces. During the summer, an improperly controlled or designed typical residential basement results in very common problems—moisture on walls and floor with the possibility of mold growth. The problem results from surface temperatures below the dewpoint temperature of the ambient air. Typical wall and floor temperatures in earth sheltered houses are shown in the adjacent table for houses controlled to 25.6°C (78°F) interior air temperature during the summer months. Whether condensation occurs, depends on the amount of moisture in the air and the air temperature immediately adjacent to the walls or floor. As the temperature of the air is lowered the relative humidity increases until the relative humidity reaches 100%. This point is called the dewpoint temperature and is the point where moisture begins to condense out of the air. Dewpoint temperature ranges for the Minneapolis area are shown in Fig. 3-25 for the summer months.

Comparison of the values in the two tables shows that the closest the tempera-

floor surface temp. (°C)

month	corner	center	air temp.
june	21.9	23.5	25.6
july	22.2	23.7	25.6
aug.	22.5	23.8	25.6
sept.	19.3	19.7	20.0

wall surface temp. (°C)

month	corner	center	air temp.
june	21.9	22.1	25.6
july	22.2	22.5	25.6
aug.	22.5	22.9	25.6
sept.	19.3	19.5	20.0

3-24 surface temperature of earth sheltered house

month	outdoor dewpoint temp range (°C)
june	9.4-15.6
july	11.7-18.9
aug.	12.8-18.9
sept.	7.2-13.3

3-25 dewpoint temperatures

ture of the wall or floor gets to the dewpoint temperature is 3.4°C (6.1°F). This is a comfortable margin; however, it would take the addition of only .0028 cu m/day (0.8 gal/day) of water to the air in the house (assuming a 350 cu m house with two air changes per day) to increase the dewpoint of the air to the point of condensation on the walls or floor at the worst summer conditions.

If dehumidification is required (or simply desired) a mechanical system utilizing a correctly sized in-room dehumidifier or a permanently mounted air conditioner/humidifier is the only option presently available. Permanent duct mounted air conditioning/dehumidification systems can also be used, however they may be more expensive than in-room units. Small in-room units have the following advantages:

- Relatively low cost.
- Portable, so they may be moved to a particular problem area.
- Dependable and efficient.
- Do not tend to subcool the air.
- Can be used only during the initial occupancy in a cast-in-place concrete structure, then resold.

Since earth sheltered housing exhibits cooling losses during the entire year, a mechanical system that lowers the air temperature to remove moisture may produce overcooling. With normal internal heat gains the effect of overcooling is somewhat reduced.

Although not presently available for residences, there are two additional methods of dehumidification which have potential for earth sheltered housing. These are dessicant dehumidification and the ice air conditioning system previously referred to under cooling sources. Dessicant dehumidification removes moisture from the air by the use of two dessicant plates that are alternately recharged (dried) by solar energy. This system does not require cooling and it appears that the operating energy requirement could be low. Dehumidification is accomplished with the ice system by using the ice air conditioning to subcool the air and solar energy to reheat it to a comfortable level.

A second method of preventing condensation problems involves increasing the surface temperature of the wall. This requires raising the interior air temperature or the addition of a small thickness of insulation on the outside surface of the wall. Such insulation increases the R-value of the wall yielding a higher surface temperature, however it reduces the cooling effect of the earth sheltered home. Because of this, raising of the interior air temperature is preferred.

Earth sheltered houses exhibit a unique characteristic concerning relative humidity and ambient air temperatures. If the home is made substantially of cast-in-place concrete (walls and floor) a great deal of water vapor will be absorbed by the interior air as the concrete cures. This process may continue for a fairly long period of time because the waterproofing that is installed on the exterior surface of the walls to protect the structure from external water infiltration also prevents the release of moisture from the curing concrete through the very same exterior surface. Hence, most of the water vapor is released to the interior of the structure. The dehumidification load due to this water vapor may be the most severe dehumidification load the structure will ever encounter. This can be dealt with by providing a small economical and portable dehumidifier for use during the initial occupancy period. A precast concrete structure will substantially reduce the dehumidification problem due to curing.

humidification

The need for mechanical humidification in an earth sheltered house is uncertain. The following maximum relative humidity levels are recommended to avoid excessive condensation on windows for an outside temperature of -29°C (-20°F).

window condition	maximum % relative humidity
double glazed	31%
triple glazed	50%

If the previously described 140 sq m house with 2 air changes per day is used for an example, then the following quantities of water must be evaporated each day to maintain the maximum recommended humidity levels:

max. % rel. hum.	cu m/day	gal/day
31%	.003 x 8	1.0
50%	.006 x 1	1.6

This example assumes that the outdoor air is completely dry (0% relative humidity) and that no internal sources of humidity are available. However, as previously indicated, there are several internal sources of humidity available. It appears probable that the sources of internal humidity will meet the small requirements calculated. In fact they may cause excessive internal winter humidity conditions. If, however, these sources are insufficient to maintain the desired humidity levels a small portable or duct mounted humidifier can be installed. If excessive winter humidity levels are encountered the outdoor ventilation can be increased.

hvac controls

Regardless of the degree of complexity of the HVAC (Heating-Ventilating-Air Conditioning) in a building, its operating efficiency is to some degree governed by how it is controlled. Although most heating/cooling systems for earth sheltered houses can be controlled by simple thermostats, it is important to examine the house and its heating/cooling systems to determine the best thermostat operation.

An important consideration for earth sheltered or any other building which has a high mass to heat loss ratio is the effect of a night setback thermostat. When the thermostat setting is lowered in the late evening, the stored heat in the mass of the house is so great and the heat loss so small that the lower temperature may not be reached by the time the temperature is raised in the morning. After the thermostat setting is raised the output of the heating system may be too small to raise the temperature of the large mass for several hours. The net effect of this may not be desirable.

In the near future it may be possible to gain energy efficiency in residential control systems through the use of solid state micro procesors (control systems with computer circuitry). These control units may be able to measure and control temperature, relative humidity, oxygen levels, odor levels, and the amount of intake and exhaust air. With this information, the control unit could determine the most efficient operation of the heating/cooling and ventilation systems.

energy analysis

The energy saving advantages of earth sheltered housing are illustrated in this section by comparing the net overall energy consumption of various earth sheltered and above grade structures. Drawings of these examples and summaries of their energy use are presented throughout this section. This is followed by discussions of the impact of shape, orientation, and additional window openings on energy use.

The assumptions on which the following data is based and the methods of calculation are presented in Appendices A, B and C as well as in some of the preceeding discussions on energy use. Several points about these calculations should be emphasized. First, these seasonal load values are based on the average monthly performance of the structures. As such, the calculated loads provide a valid estimate of the seasonal energy cost for each structure and a basis of comparison among the various cases. However, as daily peak loading factors due to extreme weather conditions have been lost in this averaging process, these results should not be construed to represent a design value for sizing the HVAC systems. If temperature fluctuations simulating real weather conditions had been included in the calculations, the earth sheltered structure would compare even more favorably to the above grade house since these rapid changes in temperature do not have as great an impact on a high mass structure.

A second important point about the following data is that the transmission losses and gains actually include heat losses and gains due to solar radiation as well. Thus the orientation of window openings, which are primarily responsible for solar radiation gains, affect the overall transmission figures. It is further assumed that all of the available passive solar energy is absorbed and used in the house with none of it being reradiated to the outside. It is also important to note that the calculations for the earth sheltered examples are based on their projected performance after three years. During the initial period after construction additional energy is required to heat the surrounding earth mass until a steady state condition is reached. Another point about the figures concerns the internal heat and ventilation loads which are used. These are considered to be reasonable values but such loads can vary somewhat depending on many factors. Finally, in reading the energy use summaries, note that a negative sign indicates heat flowing out.

case A: fully earth sheltered—one level

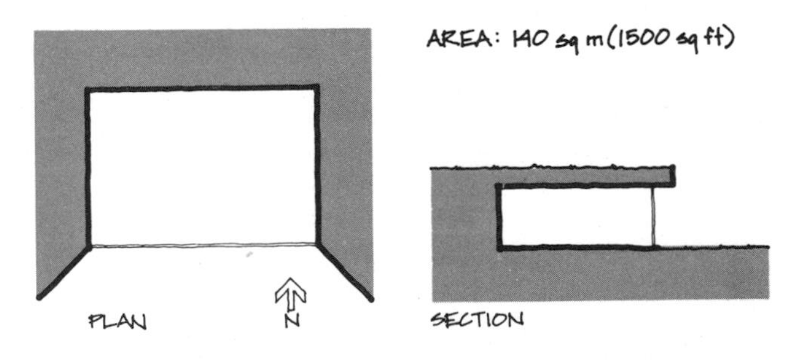

	winter (kW-hr)	summer (kW-hr)
transmission	−4043	− 969
ventilation	−1544	—
internal heat	+5900	+1700
net energy	**+ 313**	**+ 731**

case B: above grade house on slab

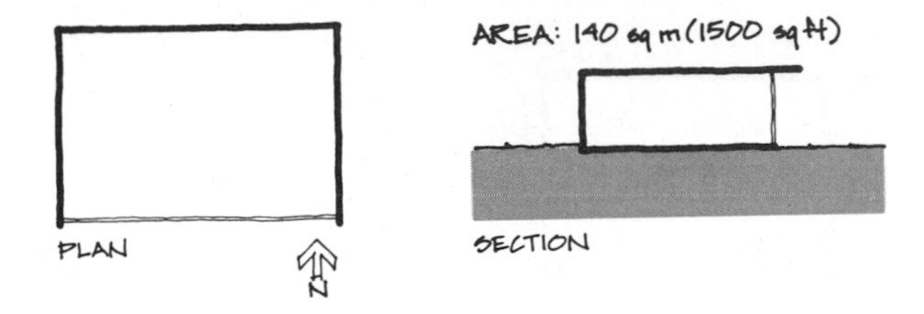

B-1

	winter (kW-hr)	summer (kW-hr)
transmission	−12935	+3658
ventilation	− 1544	—
internal heat	+ 5900	+1700
net energy	− **8579**	**+5358**

B-2

	winter (kW-hr)	summer (kW-hr)
transmission	−10186	+3173
ventilation	− 1544	—
internal heat	+ 5900	+1700
net energy	− **5830**	**+4873**

B-3

	winter (kW-hr)	summer (kW-hr)
transmission	−13697	+3002
ventilation	− 1544	—
internal heat	+ 5900	+1700
net energy	− **9341**	**+4702**

B-4

	winter (kW-hr)	summer (kW-hr)
transmission	−10945	+2519
ventilation	− 1544	—
internal heat	+ 5900	+1700
net energy	− **6589**	**+4219**

The first two cases to be compared are a fully earth sheltered, one level elevational structure (Case A) to a conventional above ground slab-on-grade house (Case B). Both structures contain 140 sq m (1507 sq ft), measure 14 m (46 ft) on the north and south facing walls, 10 m (33 ft) on the east and west facing walls, and have a ceiling height of 2.5 m (8.2 ft). The earth sheltered structure (Case A) has earth placed against the north, east and west walls. The exposed south facing elevation has 35% window area. The roof consists of 50 cm (20 in) of soil placed over 10 cm (4 in) of polystyrene insulation on a 30 cm (12 in) precast concrete plank. The insulation is also placed on the upper part of the walls to a depth of 2.15 m (7.1 ft) below the surface of the soil.

Four variations of the surface house (Case B) are presented with differing window arrangements and amounts of insulation. The first two variations (Cases B-1 and B-2) have exactly the same window arrangement as the earth sheltered house (35% on the south wall with no windows on the remaining walls). Thus, these cases can be directly compared to Case A and the results are not distorted by differing amounts of solar radiation through the windows. Case B-1 represents a house with a standard amount of insulation (type X in Appendix B). In the last two variations of the above grade house, (Cases B-3 and B-4) windows are placed on all four outside walls since this represents a more typical arrangement. These cases have the same total area of glass as Case A but it is distributed in the following manner: 15% of the south wall, 10% of the east and west walls, and 5% of the north wall. The dimensions and shading factors for these windows are given in Appendix B. Again, Case B-3 represents a house with standard insulation and Case B-4 is well insulated.

Examining the results listed in Cases A and B, it is evident that the transmission loss of the earth sheltered home is far less than any of the above grade models. If the structures are assumed to be tightly constructed so that uncontrolled infiltration is eliminated, the addition of the ventilation load and internal heat generation to the transmission losses results in a surplus of energy (313 kW-hr) for Case A during the winter. This means that the heat gained from passive solar collection and internal heat sources is greater than the loss through the building envelope. During the same seven winter months all of the Case B variations have net energy deficits ranging from 5830 kW-hr to 9341 kW-hr. This represents the energy required to maintain an interior air temperature of 20°C (68°F).

During the summer months, the surrounding earth continues to cool the earth sheltered home, although this rate is reduced somewhat from that of the average wintertime cooling. If the interior air temperature is allowed to rise to 25.6°C (78°F) in the summer, the thermal mass of the surrounding soil will draw 969 kW-hr of heat energy from the earth sheltered house. In the same three summer months, the conventional structure is being heated by the sun and outside air. This results in heat gains ranging from 2519 to 3658 kW-hr for the above grade models in addition to the internal heat load. Examining the total summertime energy gains, it is evident that the above grade variations have cooling loads six to eight times greater than the earth sheltered house.

Another single level model was considered which is a partially earth sheltered house as shown in Case C. This particular example is quite similar to Case A in that earth is placed against the walls of the structure except for an exposed south elevation. As in Case A, 35% of the south wall is window area. The only difference between the two cases is that Case C has no earth cover on the roof. It is apparent that such a modification is of little consequence to the satisfactory wintertime performance of the home. However, the additional solar input during the summer causes considerable heating of the roof structure resulting in a 2326 kW-hr heat load. While this is a less desirable situation than for the fully earth sheltered structure (731 kW-hr/summer), it is still about half of the conventional slab on grade heat load.

Even though it is unsuitable for housing, a completely below grade chamber with no window openings is shown as Case D. This case is presented for comparison purposes only so that the characteristics of earth sheltered space in a pure form can be illustrated. Note that the transmission loss in the winter is actually higher than that of Case A which has the south wall exposed. Even though the heat loss through the south wall in Case A is greater than through the earth in Case D, the passive solar heat gain through the windows (which is included in the transmission calculations) more than offsets the loss. Also note that the energy use in summer is considerably better than any of the preceeding cases.

Extending the benefits of reduced wintertime heat loss and natural cooling during the summer to a fully earth sheltered two level home, Case E shows that the plan area of the home can be doubled without incurring an energy penalty. This remarkable performance is due to two major factors. The first is that a two level plan (Case E) represents a far more compact configuration than a single

case C: partially earth sheltered

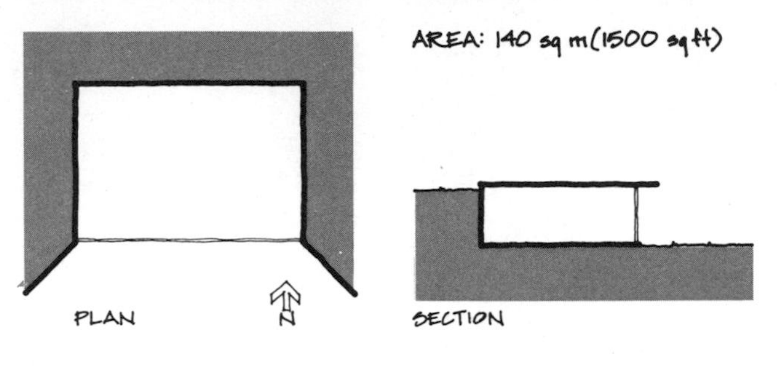

	winter (kW-hr)	summer (kW-hr)
transmission	−4244	+ 626
ventilation	−1544	—
internal heat	+5900	+1700
net energy	+ 112	+2326

case D: subgrade chamber

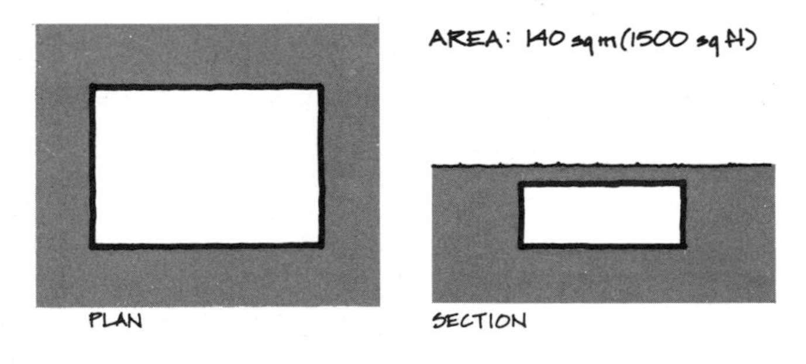

	winter (kW-hr)	summer (kW-hr)
transmission	−5159	−2734
ventilation	−1544	—
internal heat	+5900	+1700
net energy	− 803	−1034

case E: fully earth sheltered—two level

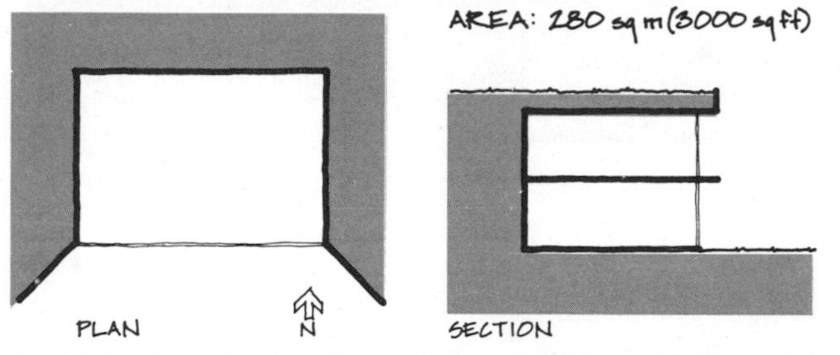

	winter (kW-hr)	summer (kW-hr)
transmission	−3884	− 634
ventilation	−3087	—
internal heat	+6600	+1800
net energy	− 371	+1166

case F: above grade house with basement

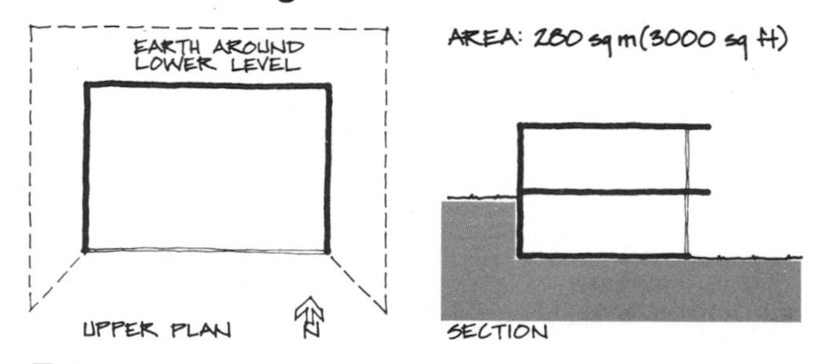

F-1

	winter (kW-hr)	summer (kW-hr)
transmission	−9876	+3358
ventilation	−3087	. —
internal heat	+6600	+1800
net energy	−6363	+5158

F-2

	winter (kW-hr)	summer (kW-hr)
transmission	−7136	+2872
ventilation	−3087	—
internal heat	+6600	+1800
net energy	−3623	+4672

level plan (Case A). In fact, while Case E contains twice the floor area of Case A, it only has 30% more total surface area (520 sq m vs. 400 sq m). Since heat loss is a function of the area through which the heat can be transmitted, it is obvious that minimizing the surface area of a structure by using a more compact configuration has significant benefits. The second contributing factor to the superior performance of Case E over Case A is that the deeper second level walls are better insulated from seasonal fluctuations. Thus despite increased ventilation loads and internal heat generation, the building still retains a low winter heating requirement of 371 kW-hr and a relatively low summer heat gain of 1166 kW-hr. It should be noted that, as with Case A, 35% of the south wall area on both levels is windows.

In northern climates a frequently employed two level configuration is the conventional above grade home with a walkout basement, as shown in Case F. The two variations of this above grade house represent a model with standard insulation (Case F-1) and a well insulated model (Case F-1). These are based on construction types X and Y discussed in Appendix B. For both cases, the window area is 35% of the south wall on both levels with no windows on the remaining three sides. From the figure it is apparent that while the below grade section of the home considerably improves the performance over a slab on grade structure, the above ground portion so handicaps the building that, upon comparison, it falls far short of the performance of the fully earth sheltered two level example (Case E).

An important issue related to the overall energy performance of houses is the rate at which the inside temperature will drop in the event of a power failure. This is of special concern to areas such as Minnesota where blizzard conditions can result in protracted power outages during severly cold conditions resulting in a rapid drop in building temperatures thus endangering both life and property. Taking the maximum rate of heat loss for test case A, which occurs in January, and providing no allowance for any internal heat gain, such as from the people inside or a wood burning stove, the temperature of the building shell will drop at a rate of less than 1°C (1.8°F) per day. Thus, not only is a large measure of safety provided, but such conditions could be prolonged for several days before the structure would begin to become uncomfortable. These conclusions agree with the experiences of current owners of earth sheltered housing (see Davis house, Part B).

impact of orientation

As previously mentioned, the transmission gains and losses in the preceding cases include the heat gains due to solar radiation. Since all of the cases have south facing windows, the overall energy performance is enhanced by this passive solar collection. In order to illustrate the impact of proper window orientation, the earth sheltered house (Case A) is shown here with the exposed window wall facing in three different directions. In all cases, the window area is 35% of that of the exposed wall.

As shown in Case A-3, where the window wall is facing directly east or west, a significant increase occurs in the transmission loss in the winter, when compared to Case A (7135 kW-hr vs. 4043 KW-hr). In addition, the east/west orientation has a dramatic impact on the summer transmission loss reducing it from 969 kW-hr to 243 kW-hr. The cooling effect of the surrounding earth walls is diminished because of the great amount of direct solar radiation on the east or west wall. In the summer, much of the solar radiation does not reach the south wall since the sun is at a higher altitude and the wall is shaded by an overhang. Proper shading on the east and west wall can help to reduce this heat gain.

When compared to the south orientation (Case A), facing the window wall to the north as shown in Case A-4 results in a very significant increase in the transmission losses (10,107 kW-hr vs. 4043 kW-hr). Since there is no solar radiation from the north, the windows become a much more negative factor in the overall energy use during the winter. On the other hand, the elimination of solar radiation actually increases the summertime cooling effect from 969 kW-hr to 1105 kW-hr. Obviously, the three cases shown here represent the most extreme conditions. Thus, orientations which fall between these cases will produce more moderate results.

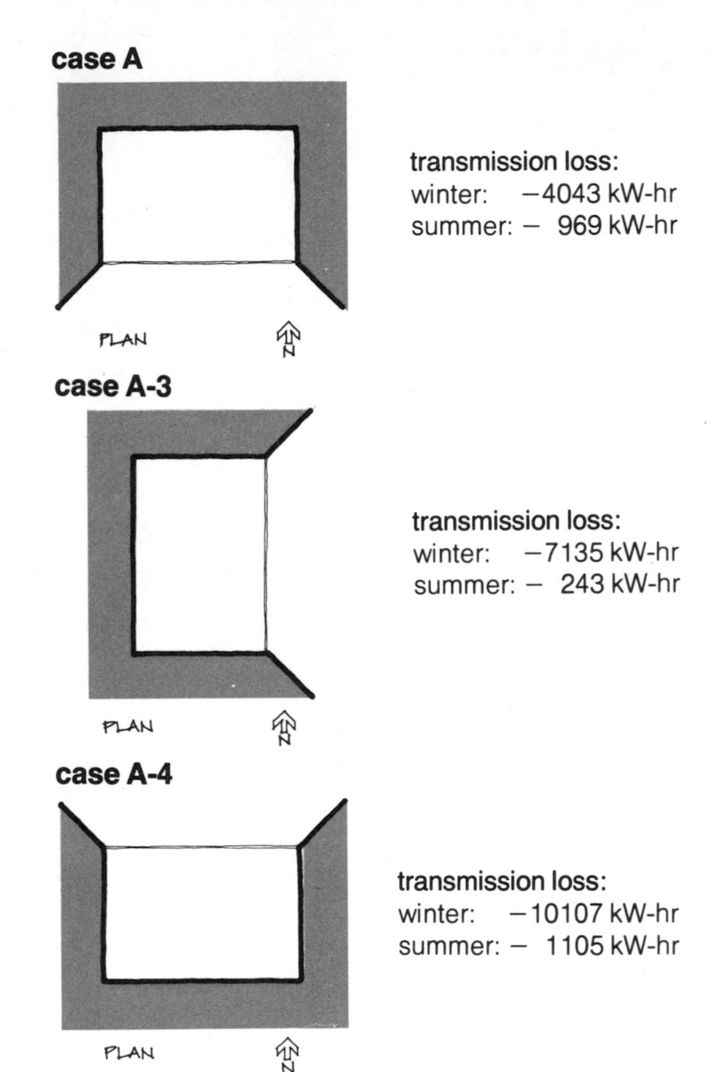

impact of shape

The primary motivation for questioning shape variations stems from the realization that where it is possible to place all the windows along a single wall, the configuration will frequently call for a longer, thinner structure than the 10 m x 14 m format examined in the previous cases. For the sake of comparison, therefore, a second floor plan (A-2) was developed for an earth sheltered home in which

case A

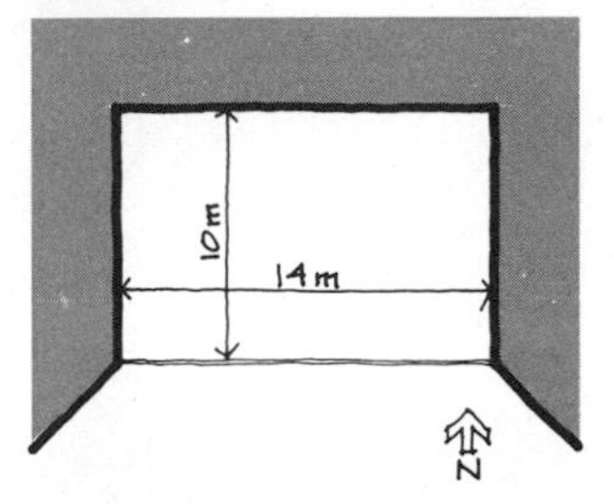

transmission loss:
winter: −4043 kW-hr
summer: − 969 kW-hr

case A-2

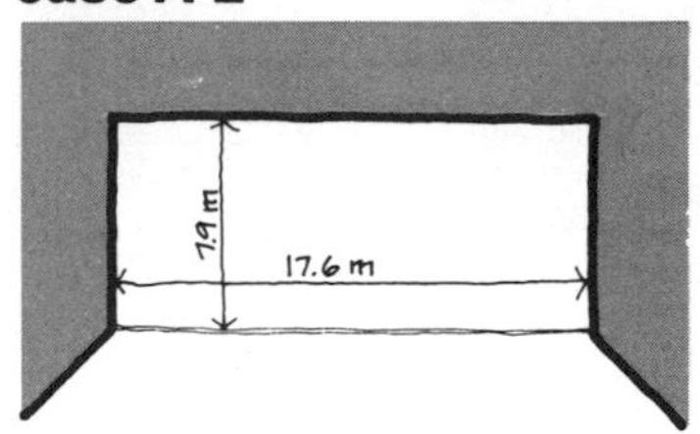

transmission loss:
winter: −4094 kW-hr
summer: − 822 kW-hr

case A-5

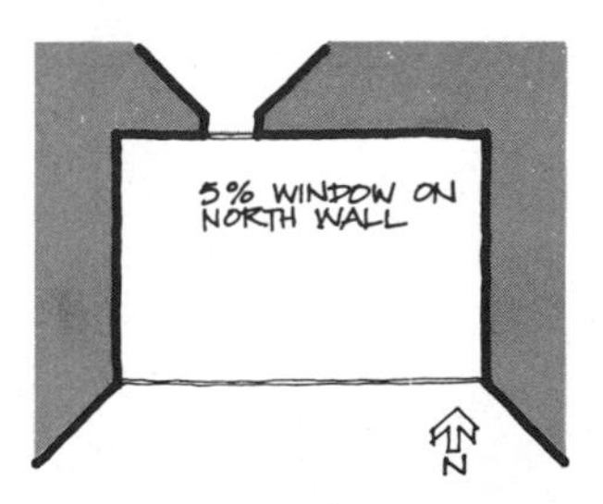

transmission loss:
winter: −4479 kW-hr
summer: − 824 kW-hr

case A-6

10% WINDOW ON NORTH WALL

transmission loss:
winter: −4916 kW-hr
summer: − 680 kW-hr

the plan area remained constant at 140 sq m (1500 sq ft), but the south and north walls were now 17.6 m (57.7 ft) in length while the east and west walls measured 7.9 m (26 ft). The net result was a 51 kW-hr increase in winter heat losses and a 147 kW-hr decrease in summer cooling due to the higher proportion of exposure on the south wall. This degree of shape variation is acceptable since the overall energy performance is still quite favorable when compared to an above grade house. Both of these cases (A and A-2) represent relatively simple, compact building envelopes which is one contributing factor to their energy efficiency. As more complex configurations are developed, thus increasing the exterior surface area, more heat will be lost through the structure.

impact of window openings

In the preceding analysis of earth sheltered structures, windows were omitted from all but the exposed south walls of the various cases. This was done to increase the collection of passive solar energy and to maximize the benefits of the earth cover on the remaining three walls of the house. Since this simple plan arrangement is not always suitable or desirable, it is important to clarify the energy use impact of additional window openings. A comparison was made between Case A from the previous analysis and four cases with various additional window openings into the earth covered walls.

In this comparison, the wall heat loss information from the numerical solution for a one level subgrade elevational structure was combined with the window performance data gained from the conventional above grade analysis. The insertion of window openings not only eliminates that amount of earth cover insulation but also increases the surface area facing the outside weather and brings this surface to the wall. Thus, a deterioration of the effectiveness of the remaining soil cover can be anticipated. This latter effect is not included in the values presented here, therefore this data should be regarded as conservative in nature in that it somewhat understates the impact of the window openings.

As shown in case A-5, the replacement of only 5% of the north wall with windows increases the winter heat loss by 436 kW-hr which is 10% of the loss for the entire house. In addition, the cooling effect of the walls in the summer is reduced by 15% or 145 kW-hr. Increasing the window area on the north wall to 10% as shown in case A-6 causes further energy losses which are a significant part of the total transmission losses for the house.

When compared to the north wall, the effect of window openings on the east or west wall is quite different due to the increased exposure to direct solar radiation. The additional winter heat loss attributed to these windows as shown in Cases A-7 and A-8 is less of a constraint than their impact on summer cooling. With 5% window area (Case A-7) the summer cooling effect of the walls is diminished by 23%, while 10% window area (Case A-8) reduces it by a significant 45% or 446 kW-hr. Of course, these figures can be reduced with proper shading techniques.

It is apparent that where it is necessary to introduce natural lighting and outside view by the addition of a window opening in a non-south facing wall, it should be done with great care and discretion. A possible alternative to window openings is the use of skylights. However, skylights may result in even greater energy losses unless properly designed. While it appears that window openings in north, east and west walls diminish the energy performance of a house, they may greatly enhance the quality and acceptability of the interior spaces. Thus, in some cases the energy loss attributed to them may be justifiable.

case A-7

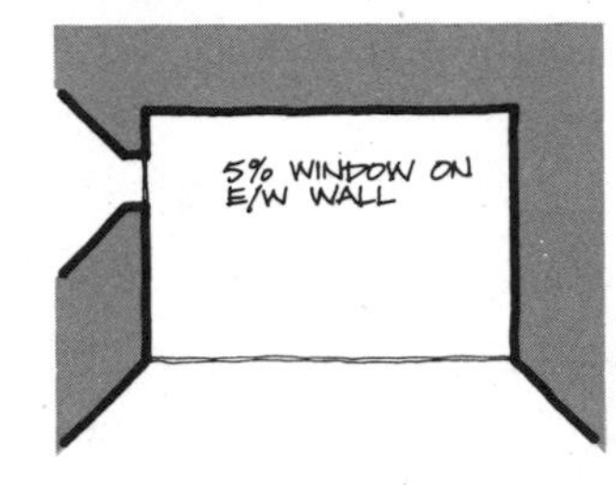

transmission loss:
winter: −4209 kW-hr
summer: − 726 kW-hr

case A-8

transmission loss:
winter: −4375 kW-hr
summer: − 523 kW-hr

4 structural design

- soils
- basic loads and structural elements
- materials
- major structural components
- application to typical layouts
- special problems and techniques

introduction

The greatest advantage of earth sheltered design will result when the necessary structure of the house is considered in the planning and layout. This aspect of design has become so commonplace and the restrictions on design so well known for conventional house structures, that they do not usually have to be a conscious part of the planning exercise. For the above ground portion of a conventional house, the loadings are relatively light and hence the load paths by which a load (such as from wind, or snow) are carried to the ground are not crucial to the design. The conventional structure can be modified easily according to well known empirical rules, for example, doubling headers and studs around windows or where large holes are cut in the floor. For a conventional basement, the burial depths are generally less than 5-6 feet of earth and for these depths of burial it has been found by wide experience that plain concrete, concrete block or treated wood foundations are, in the overwhelming majority of cases, quite adequate structurally. Because the burial depths (and hence the earth pressures) are low, there is little to be saved in the structure by determining the exact soil type and from that a better estimate of the actual earth pressure that will act on the wall.

In contrast to the conventional house, the earth sheltered house has large depths of burial, the earth on the roof is a very significant load (approximately l00-l20 lbs/square foot of roof for each foot depth of earth cover) and the structural form of the house can have a great impact on the ease with which these heavy loads can be carried. In some cases, as with arch structures for instance, the structural form will dominate the entire layout of the house. In more standard methods of construction there is a fair amount of flexibility in the design but the needs of the structural system must be borne in mind or a price will be paid in higher than necessary structural costs. Because of the heavier loads, openings in structural members must be made with more care.

This section cannot instruct a non-engineer how to design and size structural systems for an earth sheltered house. There are many variables in the determination of earth loads and pressures and the design of structural members carrying such heavy loadings should not be attempted by an amateur. If standardized designs were prepared for particular elements of the structure for different earth loads on the roof and depth of burial on the walls, they would necessarily assume the worst earth loading conditions likely. For more favorable earth loading conditions which would normally occur, the structure would be overdesigned. This doesn't matter much for lightly loaded structures (e.g. the conventional basement) but as the earth loads and the structural strength required increase, larger and larger savings can normally be made by a careful individual structural design.

It is recommended here that the services of a professional engineer (usually listed in the Yellow Pages of the telephone directory under consulting engineers) be employed to prepare or at least check the structural design of an earth sheltered house. In some cases this will be done independently of the future owner. For instance, if an architect is involved he will arrange for any necessary assistance in this area. If precast concrete plank manufacturers are involved, they often have registered professional engineers on their staff who prepare the structural design for the parts of the structure where their planks are used. Reputable contractors will usually have an engineer who they can turn to for the design of structures outside their normal competence. This individual design approach to earth sheltered structures may become less necessary as more such structures are built in a particular area, particularly if the same firm is involved in several houses. For the design of similar elements in this case, it would only be necessary to check the site investigation information to make sure that the loading and support criteria were the same in order to determine whether the same structural design could be used.

It should not be construed here that earth sheltered housing only gives problems with the structure. There are in fact several advantages to offset the greater strength and more involved design required. Some of these advantages are listed below:

1. The structures will generally be constructed of massive and relatively permanent materials which will give the buildings a very long expected life.

2. The buried structure will be subject to much smaller temperature fluctuations than an above-ground structure. For a conventional roof for example, the temperature of the roof surface may fluctuate in Minnesota from well below

0° F in the winter to over I40° F on a hot sunny day in summer. For a buried roof, especially under the insulation recommended, the temperature fluctuation of the structure will be only of the order of 5° — I0° F.

3. Both the advantages above will contribute to a low maintenance structure.

4. Concrete construction methods will give a very fire resistant structure. Should a fire occur, the structure is likely to be little damaged by the fire unless an unusual fire hazard from stored materials exists in the house.

5. Simple layouts and structural systems (which reduce costs) can be used because the attractiveness of the house from the outside may not be as dependent on an interesting (and usually expensive) exterior shape or on the use of expensive facing materials for all the walls and roof. In contrast to a conventional above grade house, the aesthetic appeal of the design is likely to depend more on landscaping and the way the house blends with its surroundings.

It is the intent of this section to identify the major parameters affecting the structural design of an earth sheltered house so that a house can be conceptually designed without making layout decisions that will cause expensive solutions in the structural design. The cost for the structural design process will be kept to a minimum if the house has a sound layout before it is brought to the structural engineer. In this case, he will only have to design the necessary structural members and provide the details and any specifications, etc. If the house is poorly layed out when he becomes involved, several stages of redesign may be involved before an acceptable compromise between structural system cost and desired layout is achieved. In keeping with this intent, most of the discussions are general in nature with advantages and disadvantages of various systems given and special problems noted.

classification of soils

The type of ground in which the house excavation is made will have several implications for the design. For geotechnical engineering purposes soils are generally classified using the soils classification charts shown on this and the following page. This will clarify the type of soil and in addition some information on the compactness (or stiffness in the case of clay) of soil is included in a soils engineer's description. Strength and loading parameters of the soil are usually estimated for small structures such as a house from the soil description and use of simple empirical measures such as a penetration test which involves forcing a probe into the soil and measuring the force or number of blows needed to move the probe a certain distance.

Rock is classified somewhat differently since the material itself is usually very strong but the overall strength of an excavation wall etc. is dependent on the jointing of the rock. Except for certain prime locations such a river bluff, the cost of small excavations in rock will generally make houses dug into rock an economically unattractive proposition.

implications of soil type for structural components

roof

Most soils are close enough in density that any density variations will not have a large impact on the house design. The best estimate for soil density should, of course, be used in the actual structural design. The strength of the soil will not generally be important for the roof design except in curved roofs or where the depth of soil on the roof approaches twice the span of the roof (an unlikely condition). The soil on the roof is usually treated as a permanent load equal to its total weight and the roof must carry this entire load. The susceptibility of the soil to frost heave and the ability of the soil to support vegetation are the major characteristics to ascertain. Fine grained soils, especially silts, are the most susceptible to frost heave . For small depths of burial the roof will not require large amounts of fill and it may be possible to bring a suitable material to the site for the roof cover if the site material is not suitable.

<table>
<tr><th colspan="3">major divisions</th><th>letter symbol</th><th>typical descriptions</th></tr>
<tr><td rowspan="8">coarse grained soils

more than 50% of material is larger than no. 200 sieve size</td><td rowspan="4">gravel and gravelly soils

more than 50% of coarse fraction retained on no. 4 sieve</td><td rowspan="2">clean gravels
(little or no fines)</td><td>GW</td><td>well-graded gravels, gravel-sand mixtures, little or no fines</td></tr>
<tr><td>GP</td><td>poorly-graded gravels, gravel-sand mixtures, little or no fines</td></tr>
<tr><td rowspan="2">gravels with fines
(appreciable amount of fines)</td><td>GM</td><td>silty gravels, gravel-sand-silt mixtures</td></tr>
<tr><td>GC</td><td>clayey gravels, gravel-sand-clay mixtures</td></tr>
<tr><td rowspan="4">sand and sandy soils

more than 50% of coarse fraction passing no. 4 sieve</td><td rowspan="2">clean sand
(little or no fines)</td><td>SW</td><td>well-graded sands, gravelly sands, little or no fines</td></tr>
<tr><td>SP</td><td>poorly-graded sands, gravelly sands, little or no fines</td></tr>
<tr><td rowspan="2">sands with fines
(appreciable amount of fines)</td><td>SM</td><td>silty sands, sand-silt mixtures</td></tr>
<tr><td>SC</td><td>clayey sands, sand-clay mixtures</td></tr>
<tr><td rowspan="6">fine grained soils

more than 50% of material is smaller than no. 200 sieve size</td><td rowspan="3">silts and clays</td><td rowspan="3">liquid limit less than 50</td><td>ML</td><td>inorganic silts and very fine sands, rock flour, silty or clayey fine sands or clayey silts with slight plasticity</td></tr>
<tr><td>CL</td><td>inorganic clays of low to medium plasticity, gravelly clays, sandy clays, silty clays, lean clays</td></tr>
<tr><td>OL</td><td>organic silts and organic silty clays of low plasticity</td></tr>
<tr><td rowspan="3">silts and clays</td><td rowspan="3">liquid limit greater than 50</td><td>MH</td><td>inorganic silts, micaceous or diatomaceous fine sand or silty soils</td></tr>
<tr><td>CH</td><td>inorganic clays of high plasticity, fat clays</td></tr>
<tr><td>OH</td><td>organic clays of medium to high plasticity, organic silts</td></tr>
<tr><td colspan="3">highly organic soils</td><td>PT</td><td>peat, humus, swamp soils with high organic contents</td></tr>
</table>

4-2 soil classification chart
dual symbols used to denote borderline classifications

walls

There are larger differences in the lateral pressures on walls between different soil materials. Most soil materials can still be used and it is more important in this case because of the greater quantities required to use as much of the local site material as possible for landscaping and backfill. It is expensive to haul material away from the site and to bring in new material. Nevertheless, when considering the structural design of the walls in conjunction with the drainage and waterproofing requirements, it is customary to use a sand or gravel backfill immediately adjacent to the wall which will drain groundwater to the foundation drain. The major type of soil to avoid is an expansive clay. Expansive clays swell when they become wet and can exert high pressures which are too large to economically design against. Soft clays at the site can also present problems of excavation slope stability and difficult site conditions during construction.

foundations

The major parameter here is the permissible bearing capacity of the soil. For the walls, which carry a substantial portion of earth loads from the roof the foundation loads will be much higher than for a conventional structure. This will make a good estimate of the bearing strength of the soil worthwhile since general code values are necessarily quite low where no special determination is made (see site investigation procedures). Soils to avoid at the foundation level are any soft or loose deposits. Special care should be taken when dealing with existing man-made fills, because very uneven bearing conditions may exist which can cause severe structural problems. A properly compacted fill, however, should not cause problems.

floors

The floor of the structure will not present any structural problems under normal circumstances. Extra problems will have to be dealt with when:

1. A high water table is not drained and a water pressure will act on the underside of the floor.
2. Expansive clays are present.
3. Deep excavations are made in soft clays. A house structure would not

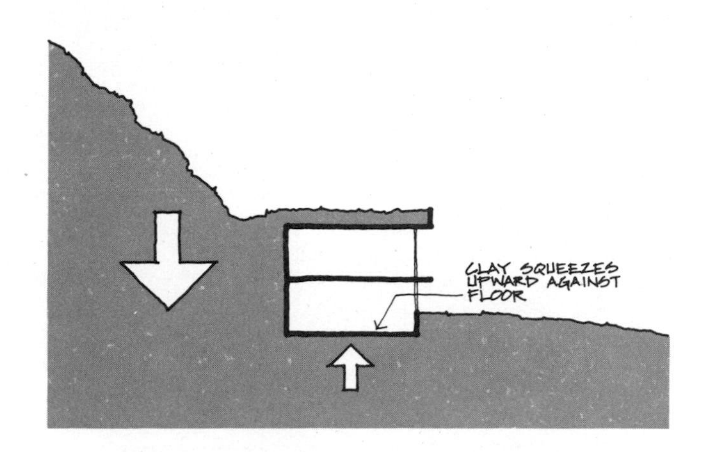

4-3 floor uplift loads

normally be deep enough to cause problems except when a hill exists on one side of the structure increasing the vertical load at the side of the excavation which can cause squeezing of the clay against the underside of the floor as shown in the adjacent drawing.

suitability of various soils

The following list indicates the appropriateness of various types of soil for an earth sheltered house. Most soils will not fall neatly into these categories, but a more detailed discussion is not possible without detailed data from a soil investigation.

category	type	qualifiers	suitability
cohesionless	gravels sands	very loose loose	good drainage but may need compaction for adequate bearing.
		med. dense dense very dense	excellent—good drainage, good bearing, low lateral pressures
	silty sands clayey sands		will depend on whether cohesive or cohesionless elements dominate its behavior. should generally be workable unless soft or loose conditions prevail.
cohesive	silts clays	very soft soft	careful evaluation needed.
		med. stiff stiff very stiff hard	should present no particular problems structurally. drainage of high water requires granular backfill. septic tank system can have problems.
		expansive	avoid
	highly organic soils	peat humus swamp soils	would probably require extensive replacement of soil or use of special foundation techniques.

4-4 soil suitability

site investigation procedures

In the absence of other advice from a competent architect or engineer familiar with local conditions, the following information should be sought for use in house design and in determining the suitability of a site. Soil testing firms usually carry out the physical aspects of the work and they, or a consulting engineer, can supply the necessary other information and recommendations.

1. At least one soil boring extending to a depth of approximately 10 feet — 20 feet below the proposed foundation unless bedrock is encountered or knowledge of local conditions make this depth unnecessary. Borings should be made using the standard penetration, split spoon method in accordance with ASTM D 1586.
2. The types of soil should be logged as the boring proceeds. Soil samples should be taken and kept in air-tight jars.
3. The depth at which groundwater appears to be encountered should be noted (this is difficult to determine exactly because drilling water or mud is normally used in the drilling process).
4. If the groundwater table is close to the footing depth of the structure, a perforated plastic pipe can be installed so that the groundwater table and any fluctuations in it can be monitored. The measurements can be made quite simply with a tape and bobber.
5. A testing firm or consulting soil engineer should comment on the suitability of the site based on their information. They should recommend design values for allowable foundation loads and lateral earth pressure.
6. To keep the subsurface investigation costs low, it will help if the potential position of the house on the site and some major parameters of the design have been considered. The location of the boring, the depth of the boring and the depth of penetration tests for bearing values will depend on such information.
7. A percolation test soil boring program is usually required when a septic tank and drain field are to be used. Testing can be done at the same time as the remainder of the investigation if this will be necessary. The cost of the subsurface investigation as outlined above will be on the order of $250 — $500 without the percolation test and $350 — $700 with the percolation test.

This section of the report presents the basic loads which must be considered for an earth sheltered structure and also establishes some basic definitions of load paths and structural elements. The loads, which are presented first, can be divided into three groups: the dead loads including vertical and horizontal soil pressures, the live loads and unique load considerations for earth sheltered structures.

basic loads

dead loads

Dead loads refer to any load on a structure which may be considered permanent. This includes the weight of the structural components such as the sheeting, beams, columns, etc. as well as the covering or finish material such as plaster, waterproofing, etc. When the loads and the span of a structure are increased as they often are in earth covered designs the dead loads due to the structure become a more significant portion of the total load. For a conventional house for example, the snow load on the roof (a live load) may be 40 lbs/sq. ft. compared to a structural weight of 10 lbs/sq. ft. For an earth sheltered roof the applied load may be 200 lbs/sq. ft. with a structural dead load of more than 100 lbs/sq. ft. In addition to the structural components other loads which can be considered permanent in an earth sheltered house are the earth loads. The vertical and horizontal earth loads are discussed below.

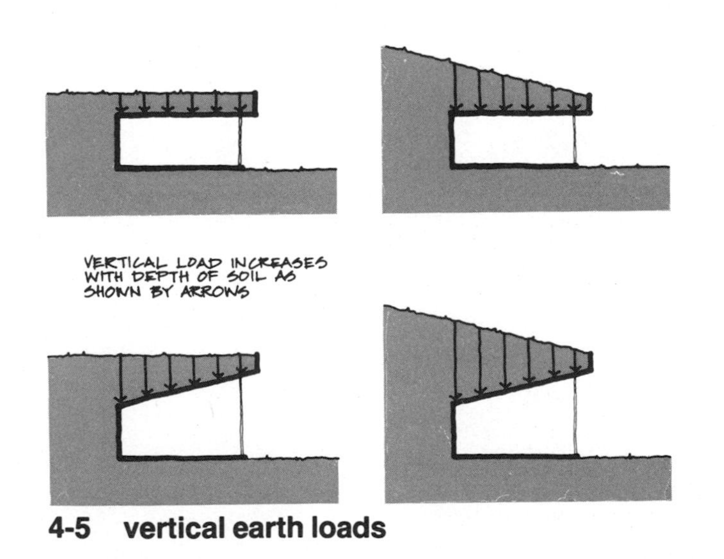

4-5 vertical earth loads

vertical earth loads

For most earth sheltered structures the full vertical load due to the soil must be carried. For most soils the load will be of the order of 100 to 120 lbs/sq. ft. for each foot depth of earth cover. For flexible structural shapes with substantial earth covering there may be a load reduction on the structure due to soil arching. It is also possible in some instances for the vertical earth load to be greater than the full overburden. The adjacent illustrations indicate the manner in which vertical loads are directly dependent on the depth of soil on the structure.

horizontal earth loads

Lateral earth pressures are determined using a pressure coefficient K. This is estimated or can be determined by tests for various soil types. A typical value for earth pressure in dry soils would be 30 lbs/sq ft lateral pressure for each foot of depth. The lateral pressure is reduced when the soil is banked or bermed against the structure and it is increased if the structure is cut into the toe of a slope as shown in the illustrations. Since variations in soil type as well as the siting of the structure itself have great impact on the horizontal loads, it is best to consult a soil engineer.

live loads

Live loads are those loads that are likely to change in magnitude and location or direction over the life of the building. They include snow, wind, furniture, people, cars, plants, etc. The distinction between dead and live loads is important in structural design because some loads are often counted on to balance other loads and also because most structural materials respond differently to short term loads as opposed to permanent loads. Most live loads are specified by the building code. The roof of an earth sheltered house presents an interesting problem. Usually the live load on the roof would be determined to accommodate people walking around. Since the roof may appear to be part of the regular ground surface, however, special care should be taken to prevent heavy trucks from inadvertently backing or driving onto the roof of the structure. Another live load imposed on the roof of an earth sheltered structure is the snow load which is also specified in the building code. In Minnesota the code values range from 30 to 50 lbs/sq ft for normal roofs depending on the location in the state. A low roof near a higher obstruction, as could well be the case for an earth sheltered home, can develop snow loads much higher than code values. Special consideration should be given to this.

A final consideration for the roof of an earth sheltered house is the loads related to vegetation. Except for trees, the main load from planting is the soil mass needed for the roots and any granular material. The weight of trees should be considered as well as the forces and movements transmitted to the structure due to the wind loads on the tree.

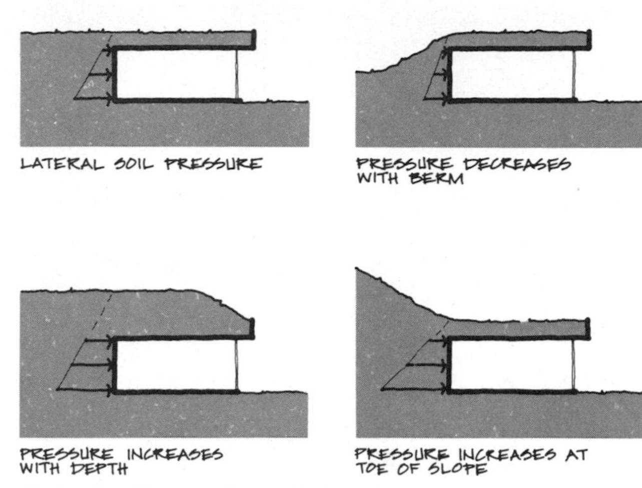

4-6 horizontal earth loads

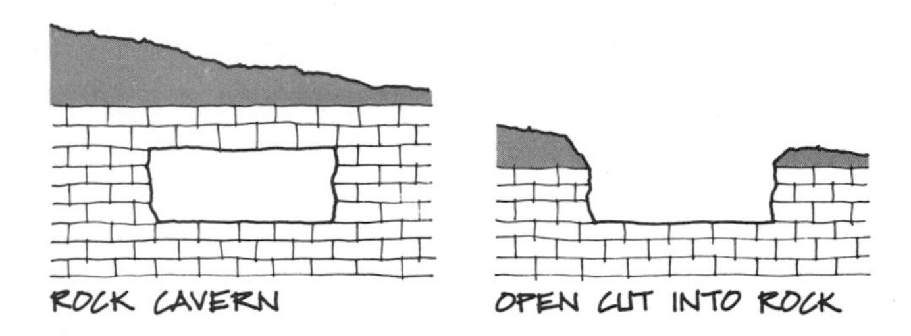

4-7 loads in rock excavations

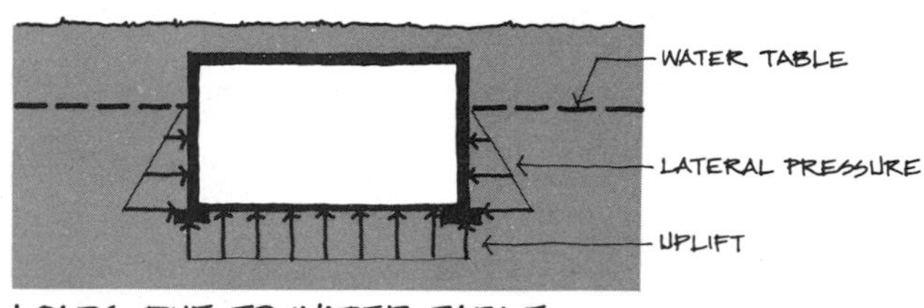

4-8 groundwater loads

4-9 frost loads

loads in rock excavations

Caverns in rock can be self supporting for certain sizes in competent rock strata. When this is true, little or no structural costs are encountered. The disintegration and haulage of rock can be expensive since drill and blast methods or very heavy rock cutting machines are needed except for very weak rock. Unless such weak rock is overlain by a harder more competent layer it would not be self supporting for usefully large openings. It is unlikely that many opportunities will exist to use a self-supporting rock roof for housing since a house must be located relatively near the surface with substantial openings. A more likely occurance would be an open cut into rock which is usually self-supporting unless there are unfavorably located joints with slippery filling. An engineer can evaluate the site for such possibilites before or during the excavation.

loads due to groundwater

Water in the ground causes extra loads on the structure. If the ground is less than saturated it must be considered as adding to the dry weight of the soil. Water of this type may increase the load on the structure. It can also cause swelling of some clays and organic soils which can result in high squeezing pressures on the structure. When the water is standing and the soil is saturated account must be taken of the extra hydrostratic pressure on the structure. The pressure exerted by water is 62.4 lbs/sq. ft. for each foot of depth. Uplift on floors can be a special problem with saturated soil. Much the same problems occur in rocks as in soil. In general, because of the waterproofing problems in addition to the increased loading that occurs when a building is below the water table, good drainage should be provided for all below grade construction.

loads due to frost

Loads due to expansion of the soil upon freezing are not well quantified. Expansion is greater for higher water contents and for silty soils. In general, this problem can be minimized by using granular backfill and providing good drainage. Parapet walls on the roof can be subject to damage from frost and should be avoided or carefully designed.

load paths

There are three types of load transfers or load paths through a structural configuration:

direct — tension and compression

The direct transfer of load requires the load to be co-axial (in-line) with the available reaction (usually supplied by the foundation). For a given length this is usually the most efficient in terms of construction material used. In compression, slender members can be subject to buckling which causes a loss in strength from that expected due to the member's cross-sectional area.

COMPRESSION

TENSION

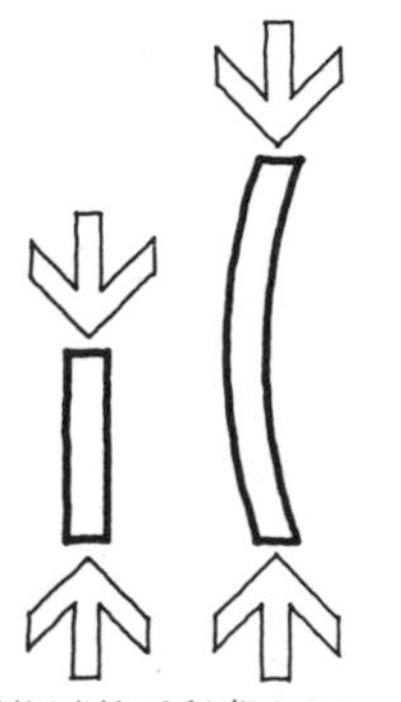

BUCKLING OCCURS IN COMPRESSION WHEN SLENDERNESS RATIO IS EXCEEDED

shear

The shear transfer of load is in simple terms the transfer of load parallel with the available reaction. The need to develop this strength between structural elements in earth sheltered houses can be easily overlooked.

SHEAR

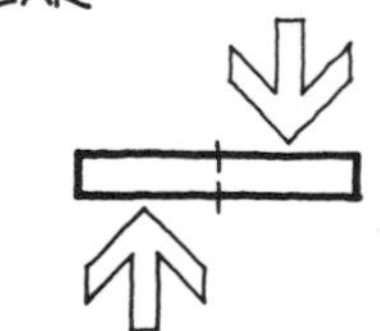

bending

The transfer of moments is often the most inefficient use of construction materials and should be minimized where possible. Bending and shear are usually present together in a structural member and which controls the design depends on shape of the member in the direction of the applied loads. Short span, deep members in the direction of the applied load introduce only low bending stresses, and hence, shear is the most important parameter. Long span, shallow members, on the other hand, will have high bending stresses which will dominate the design.

BENDING

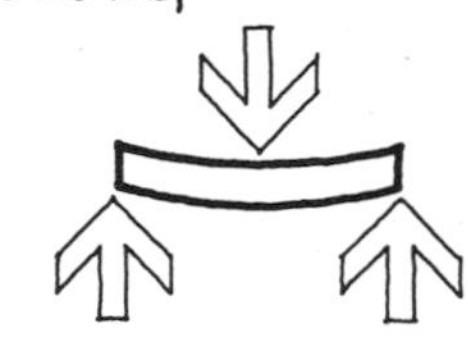

structural elements

A structure is made of many elements which can include:

Columns are usually straight elements designed to carry compressive loads by a direct load path. Most often vertical in conventional structures but often also horizontal in earth sheltered houses to transfer the large lateral earth loads. They can be wood, steel or concrete. Bracing may be needed to prevent buckling.

Ties are straight elements designed to carry tensile loads in a direct load path. Most often of steel rod or cable but sometimes wood.

Beams are usually straight elements designed to carry lateral loads by bending and shear paths. Most often horizontal in conventional houses but are also vertical in earth sheltered houses to withstand the lateral earth loads.

Beam-columns carry both direct and shearbending loads in a single member. Examples are earth covered walls and floors which brace walls.

Shear walls and diaphragms are elements carrying shear loads. They are several times deeper than long in the plane of load application, otherwise they must be treated as beams.

Plates are thin flat elements carrying loads perpendicular to their plane. These are really two-dimensional beams and can be analyzed as such unless the deflections exceed about half the thickness of the plate. If there are in-plane forces on the plate it will behave like a diaphragm. If the plate is thin it may have to be braced to prevent buckling.

Arches are elements that carry bending type loads by a direct path more efficiently than a beam would. Its curvature is hard to build and utilize. Large thrusts must be reacted at the ends of the arch which may require ties, columns and large footings.

Shells are 3-dimensional versions of arches which can carry many types of loads in an efficient direct or shear load path rather than in bending. Their curvatures are difficult to build and utilize.

Trusses are composite elements made up of beam-columns and ties and are designed to carry bending type loads more efficiently than beams. However, to do this they must be deeper than beams and take up more space. This usually restricts their use to above grade roofs.

materials

The common structural materials for earth sheltered housing are concrete, steel and wood. In this section these materials are subdivided further and their basic characteristics are presented as advantages and disadvantages. In addition, the suitable uses in an earth sheltered design of each type of material is indicated. In the following section on major structural components, there is further discussion of the basic materials with reference to each component.

plain cast-in-place concrete

uses

(1) floors on grade
(2) walls with less than about 6 feet of cover
(3) footings

advantages

(1) low cost
(2) durability, fire resistance
(3) high compressive and shear strengths
(4) rather watertight if not cracked
(5) heavy to resist uplift and sliding due to water and earth pressures
(6) can be placed in large and, or complex shapes

disadvantages

(1) low tensile strength causes cracks which leak
(2) heavy, requiring larger foundations and footings
(3) cure time can delay construction
(4) cold weather work difficult
(5) residual moisture load can cause high humidities in building for several months

reinforced cast-in-place concrete

uses

(1) floors on grade
(2) self supporting floors and roofs
(3) walls at any depth
(4) large footings
(5) columns and beams
(6) arch and vault shaped roof systems

advantages

(1) durability, fire resistance
(2) high compressive, shear and tensile strengths
(3) fairly water tight — reinforcing reduces width of cracks caused by shrinkage of the concrete on drying or structural movements.

disadvantages

(1) good structural design needed to limit cracks
(2) heavy, requiring larger foundations and footings
(3) cure time can delay construction
(4) cold weather work difficult

(4) heavy to resist uplift and sliding
(5) can be placed in large or complex shapes

(5) residual moisture load can cause high humidities for several months

precast concrete-reinforced or prestressed

uses

(1) self supporting floors and roofs
(2) walls at any depth
(3) columns and beams
(4) large culvert shapes

advantages

(1) durability, fire resistance
(2) high compressive, shear, and tensile strengths
(3) very water tight if joints done properly
(4) moderately heavy to resist uplift and sliding
(5) can be economically mass produced or standard mass produced elements can be used
(6) no cure time delay
(7) can be placed in cold weather
(8) can be built quickly
(9) less moisture load in building after construction than with cast-in-place concrete
(10) if section contains voids, these can be used for utility distribution

disadvantages

(1) good structural design needed to control opening of joints between elements
(2) heavy, requiring larger foundations and footings
(3) trucking costs high if not close to precasting plant
(4) sealing of joints
(5) shapes usually limited to linear elements
(6) long lead time for specialized shapes

reinforced and plain masonry

uses

(1) walls with less than 6 feet of cover for unreinforced concrete block
(2) walls to a maximum of 2 story embedment with reinforcing
(3) interior partitions

advantages

(1) durability, fire resistance
(2) moderate compressive, shear and tensile strengths when reinforced
(3) moderately heavy to resist uplift and sliding

disadvantages

(1) easily cracked allowing water penetration
(2) heavy requiring larger foundations and footings
(3) requires a very good water sealant

(4) elements are mass produced
(5) can be assembled into large or complex shapes
(6) a good distribution of plants to keep trucking costs low

(4) cure time can delay construction
(5) cold weather work difficult
(6) requires reinforcing for most earth sheltered uses

steel

uses

(1) beams and columns
(2) reinforcing for concrete and masonry
(3) large culvert like arches

advantages

(1) very high strength
(2) can be formed to arch shapes
(3) very water tight
(4) can be built in cold weather

disadvantages

(1) needs corrosion protection
(2) needs fire protection
(3) expensive except in certain arch applications

wood

uses

(1) self supporting floors and roofs
(2) walls to more than 1 story embedment but costs rise rapidly with increasing embedment depths
(3) interior walls
(4) on grade floors

advantages

(1) medium high strength
(2) light weight for light foundations and footings
(3) built by standard house building carpenter crews
(4) quick construction
(5) can be built in cold weather
(6) low cost

disadvantages

(1) extensive nailing or gluing needed to develop shear strength
(2) needs chemical treatment to control decomposition
(3) needs well drained earth to remain dry
(4) combustible
(5) its light weight does not resist uplift and sliding well, needs good drainage to avoid uplift.
(6) some disputes exist on the longevity of the system and long term toxicity problems with the chemical preservatives

The variety of structural systems available for earth sheltered housing can be divided into two general categories. The first group includes the more conventional systems such as concrete walls with precast or poured concrete roof structure as well as steel and wood post and beam systems. These are the most commonly used systems resulting in vertical side walls and flat or sloped roofs.

The second group consists of a variety of more unconventional structural systems including concrete and steel arch and dome shapes which have unique potential for earth sheltered structures. Since the conventional systems are the most likely to be built and have a number of general structural characteristics in common, most of the major structural components discussed in this section pertain to the first group. A brief discussion of the unconventional systems is included at the end of the section.

exterior building walls

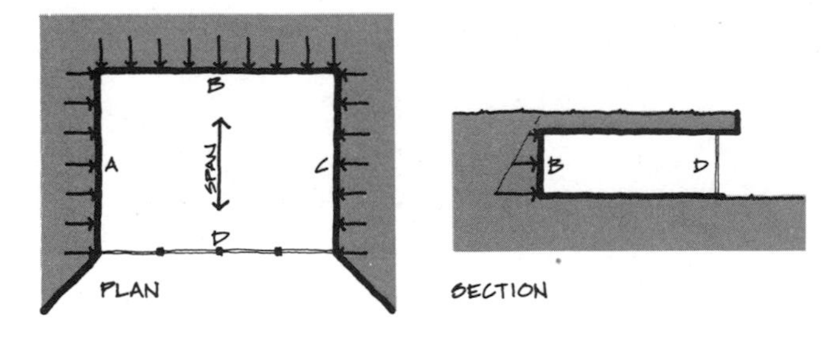

4-10 walls supporting earth loads

There are four basic imposed load conditions for the exterior walls of an earth sheltered house. Three of these can be illustrated by the simple plan and section shown at left. Walls A and C must resist lateral earth pressures but do not have to carry any substantial vertical load because the roof spans between walls B and D. Wall B must resist the lateral earth pressures as well as carry the vertical load from the earth covered roof. Wall D does not have to resist any earth pressures, but must still carry the vertical load from the roof. The fourth condition would be where the roof spanned in the opposite direction and wall D would then be free of major loads except that wind pressures must still be considered. Design winds loads are, however, small compared with earth pressures.

walls supporting earth loads

The earth retaining walls are essentially slabs loaded from the side instead of from above as in case of the roof. The walls still need to be supported and the supports must be capable of transferring the load through the structure until an equal and opposite equilbrating force can be found. Some of the typical load paths for resisting the lateral earth pressures are shown in the accompanying diagrams.

The cantilever wall is an expensive solution where floors, roof and other walls are available to provide a balancing reaction. A cantilever design has larger wall moments to be designed for and the footing must also be designed to transfer these moments to the soil.

In contrast, the balancing reactions between walls A and C induce only low compressive stresses in the roof and floor system which will not present problems in most roof or floor designs.

When the earth pressures are not balanced on either side of the house, the unbalanced reactions (e.g. between walls B and D) will require the roof to be designed as a diaphragm to transfer the load to the end walls A and C. This transfer of load in the plane of the floor can usually be accomodated without problems in a concrete roof structure. A timber roof structure cannot provide as substantial a diaphragm action although for one story designs with a small amount of earth cover sufficient shear strength should be available.

Horizontal or two-way spanning of the wall, although it avoids taking as much load up to the roof, has several drawbacks. The wall span in the vertical direction will be about 8 feet and the loading will be triangular in shape. To span the wall horizontally, substantial buttresses or partition walls would have to be provided at this sort of interval to give a competitive wall design. These could not all be incorporated into partition walls at this spacing and hence would be additional structural elements. For significant two-way action support walls would be needed at less than 12 foot intervals. Furthermore, the lowest horizontal strip of the slab must be designed for a uniform load equal to the peak of the triangular load for the vertical strip. This nessitates either an uneconomical design for the rest of the wall or changes in the design with depth.

The walls of an earth sheltered house must also in many cases act as shear walls to carry roof diaphragm forces or other similar forces down to the ground. The shear wall is, in essence, a vertical diaphragm and as mentioned previously, concrete components are usually capable of resisting these in-plane shears without any major strengthening of the structure. The design of the shear walls should only become critical if the following conditions are present:

1. long heavily loaded diaphragms in the roof or intermediate floor
2. a short effective length of shear wall (deducting for openings in the shear wall)

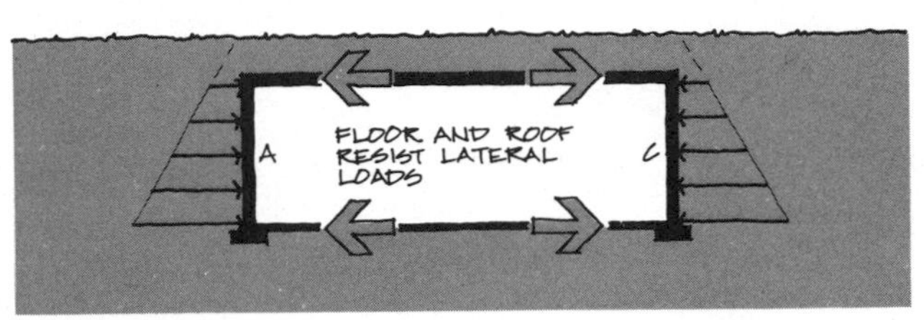

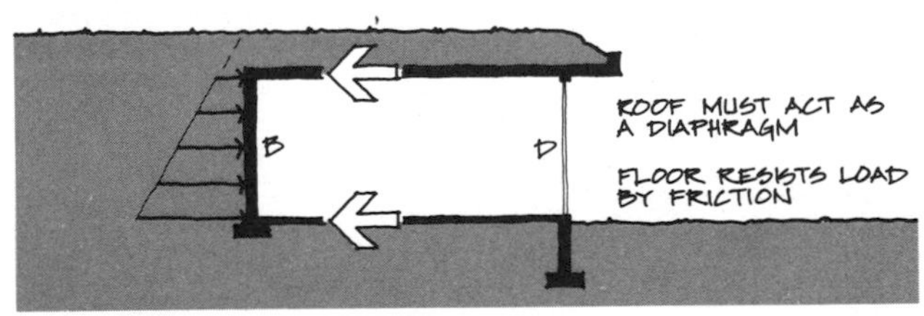

4-11 balancing earth pressures

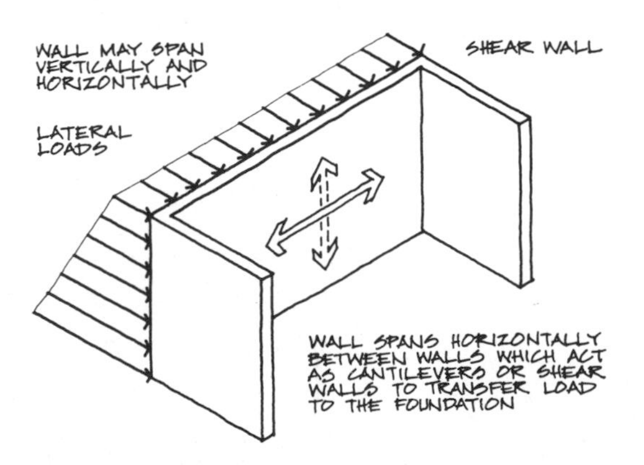

4-12 2 way wall span

If shear walls prove to be critical, the most common solution would be to add another shear wall in the middle of the structure, perhaps to replace a non-structural partition.

Walls of two level houses are also normally designed as vertical strips supported by the roof, intermediate floor, and ground slab. In this case the wall can act as a continuous beam rather than two individual spans. The maximum moment in the wall is reduced by the continuous beam action as opposed to simply supported spans but the moments are still considerably larger than for the one story case, because the lower wall section has a greatly increased earth pressure. In a two story wall, the intermediate floor provides a greater reaction to the wall than either the ground floor or the roof. This places some restrictions on the design of the intermediate floor which are discussed in a later section.

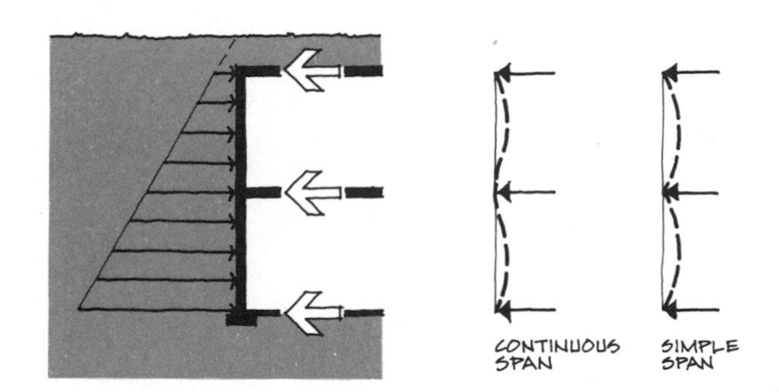

4-13 2 story walls

exposed exterior walls

These walls may carry a vertical load from the roof or they may only be required to give protection against wind and weather. Window openings will tend to be massed on any exposed wall in an earth sheltered house. Because of this, the wall cannot have the same kind of continuous load bearing capability as other walls. When the wall does support vertical loads, steel beams will usually be required above any openings. When the wall does not continuously support vertical loads it may still be required to incorporate columns in the wall to support cross beams. The question of whether these walls should be designed to carry substantial loads must be considered in connection with the roof design and the extent of the proposed openings in the wall.

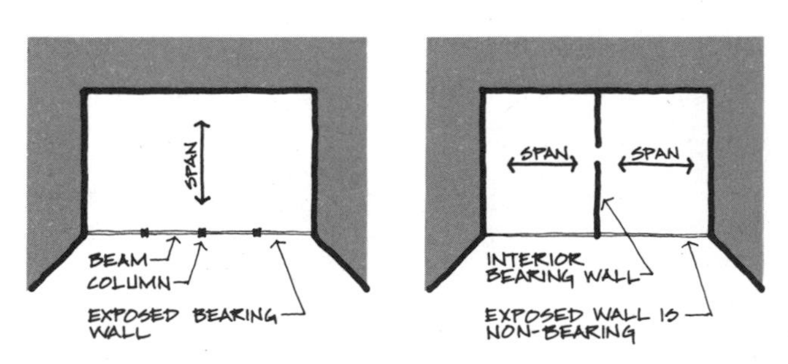

4-14 exposed exterior walls

Some design notes on the major types of wall construction are given below and are followed by a table indicating some approximate costs in the Minneapolis/St. Paul area for the different wall systems at different depths of embedment.

general

- The maximum moment in the wall when fully buried rises as the cube of the wall height. Therefore, the height of walls adjacent to the earth should be kept to a minimum for low structural costs.
- When floors and roof are designed as supports to the wall, they must be installed prior to backfilling the wall.

unreinforced block walls

- Tension in mortar joints is the critical design parameter Hence, the vertical superimposed load on the wall such as from an earth covered roof or any above ground structure, has a great effect on the design.
- No horizontal tension stresses are allowed to be counted on when a wall is laid in stack bond i.e. wall cannot span horizontally when stack bond is used. Running bond has a better structural performance.
- Type M or S mortar required below grade (see building code for mix portions).
- Allowable embedment depth for unreinforced block with no superimposed load and using Uniform Building Code stresses is just under 5 feet. With the weight of conventional one story construction above a foundation wall, this can be increased to between 5 and 6 feet i.e. conventional basement burial depth.
- Using 12 inch concrete block, a one story (8 feet story height) earth sheltered wall in unreinforced block is only feasible with a very heavy load transferred from the roof. In general, unreinforced block should be limited to use with conventional basement burial depths i.e. up to 6 feet.
- Shear walls in unreinforced block should present few design problems unless the walls are very short (less than 10 feet) or have a large total width of openings in the wall.
- The maximum allowable ratio of unsupported height or length to the thickness of the wall is 18. Hence, for 12 inch block, intersecting walls or pilasters are needed at 18 foot intervals.

reinforced block walls

- Reinforcing is included to provide the tensile resistance in bending. The reinforcing bars are placed in the voids of the wall and the voids containing reinforcing grouted solid.
- Design parameters are the size and spacing of the reinforcing bars and the compressive strength of the concrete blocks.
- 12 inch reinforced concrete block can be used for both one and two story earth retaining walls provided the intermediate floor can provide sufficient reaction. The cost rises more steeply with increasing depth of embedment than for reinforced concrete because the section is not solid and the reinforcement is not in an optimal structural position.
- The concrete blocks must be fully bedded with type M or S mortar.
- The maximum allowable ratio of unsupported height or length to the thickness of the wall is 25 to 1.
- There are minimum areas of reinforcement in both the vertical and horizontal directions - total area of reinforcement must be greater than 0.2 percent of the gross cross sectional area of the wall, minimum in either direction is 0.07 percent.
- The maximum spacing of reinforcing bars is 4 feet.

- Additional reinforcing bars are required adjacent to any openings greater than 24 inches in either direction.
- Shear walls again should not provide any design problems unless the net length of the wall is very short.

plain cast-in-place concrete

- Plain concrete basement walls should be limited to less than a full story embedment.

  Concrete intended to be watertight should have a maximum water/cement ratio of 0.48. A stronger and more waterproof concrete is obtained when a low water/cement ratio and adequate vibration of the concrete during placement is used.
- The minimum thickness of a concrete basement wall is usually 8 inches.
- Where sulphates are present in the ground in high concentrations, sulphate resisting cement should be used.
- Where pipes are embedded in concrete (e.g. for radiant heating), the temperature of the contents should not exceed 150 °F and pressures should not exceed 200 psi. Aluminum pipes should not be embedded in concrete. Pipes should not exceed 4 percent of the stress area of a structural slab without special analysis. Pipes to be embedded should be tested at 150 psi for 4 hours before concreting. This test requirement does not apply to drain pipes or pipes for Pressures of less than 1 psi above atmosphere.
- Shear wall capacity better than for concrete block.

reinforced cast-in-place concrete

- Reinforced concrete can be used for both one and two story construction.
- Reinforcement can be easily varied.
- 8 inch thick walls should be sufficient for most house construction unless the intermediate floor cannot provide the necessary reaction.
- Minimum concrete cover for reinforcement exposed to earth or weather is 2 inches for bars ¾ inch diameter or larger, 1½ inches for bars ⅝ inch diameter or less.
- Minimum concrete cover for reinforcement when concrete is cast directly against the earth is 3 inches.
- Reinforced concrete costs rise only slowly with increasing depths of embedment because only the cost of the reinforcing steel increases greatly.

precast concrete

- Precast concrete planks can be used for one and two story construction.

- The cost rises very little with increasing height of wall because although the material costs rise, bigger units are handled which reduces the handling costs per square foot of wall.
- Precast units usually have good quality control and an excellent finish. In many cases it is possible to directly spray texture walls and ceilings which saves on finishing costs.
- Structural design involving precast planks is usually done by the supplier. The supplier should either be PCI (Precast Concrete Institute) Plant Certified, have a registered Civil or Structural Engineer on their staff or obtain the stamped approval of their plans from a registered engineer.
- Generally, the more precast concrete sections used on a job from one supplier, the cheaper the cost per square foot of installed plank. If the travel distance to the job is significant, erection teams are ususally charged to the job for a full day even though installation time may only be a matter of 2-3 hours. Hence, additional planks can be installed with only a increase in material and transport costs.
- Normal plank widths vary from 2 feet to 8 feet. The narrower planks are not usually recommended for wall construction because of the higher costs in erecting and tying together the larger number of elements for a wall system.
- Precast wall panels rather than planks are also available and are generally designed for one story embedment.
- Precast concrete shear walls are designed using a shear friction approach and this should present few problems unless the shear wall is very short.

pressure-treated wood

- Wood foundation walls can be used for one story construction with widely used lumber sizes. For more than one story earth burial, member sizes increase rapidly. They appear economical in first cost for conventional basement embedment depths but quickly become uneconomical for more than one story embedment.
- Shear wall action can be a problem when the reactions from the floors are high. Nail spacing decreases rapidly as higher shear loads are carried. Again, they will normally be strong enough for one story construction with a small amount of earth cover.
- Because of connection detail problems, the wood walls are most practical when used with a wood roof system.
- When cuts in the pressure treated lumber are made on site, a concentrated preservative solution should be painted on the cut end.
- The pressure treated wood suppliers recommend a damp-proofing and drainage system to be used with the wood foundation.
- Creosote and pentachlorophenol preservatives are not permitted for wood foundations of dwellings. CCA and ACA treatments are the only waterborne preservatives permitted for this use.
- Stainless steel nails or staples are recommended except in very dry conditions where hot dip galvanized nails are a possible alternative.

depth of earth cover against wall	material				
	wood	conc. block	unreinf. conc.	reinf. conc.	precast conc.
5-6 ft.	2.45	2.70	3.75	5.40	—
1 story +2′ cover	3.55	3.20	—	5.80	4.00
2 stories to roof	7.05	3.60	—	6.45	4.20
2 stories +5′ cover	—	4.50	—	6.55	4.40

4-15 table of approximate wall costs—per sq. ft.
(minneapolis/st. paul area)

note: These prices should only be used as a general guide. local conditions may vary the relative prices greatly.

roof

The primary function of the roof is to support the vertical loads above it due to soil, snow, etc. In flat roof systems (not necessarily horizontal) this is accomplished primarily by bending. As mentioned earlier, bending is an inefficient load transfer method and care should be taken to provide sufficient vertical supports to the roof to give an economical structural system. Intermediate supports can be provided by bearing walls, beams and columns. In keeping with the general maxim that the most economical structure will result when a structural element can serve another useful purpose, bearing walls that also serve as partitions are normally preferable to support beams. Exceptions are discussed in the analysis of the basic house designs. Bending moments induced in the roof are proportional to the vertical load imposed and also the square of the span. Spans should, therefore, be kept as short as practical. Economical spans will depend on the depth of earth cover, the total dimensions of the house and the structural material used.

Choices for the roof material are mainly reinforced cast-in-place concrete, precast concrete planks or heavy timber planking and beams.

Precast concrete T-beams which are deep sections used for long span conventional roofs are not well suited for earth covered roofs because their flanges are too thin to transfer a heavy soil load across to the main spines. Precast concrete planks will generally span in one direction only across the roof. With 1½ - 2 feet of earth cover, spans in excess of 30 feet are uneconomically high for generally available planks. Maximum economical spans will probably be of the order of 25 feet. If shear is a problem, (not usually the case) the voids in the planks can often be omitted to give an increased shear strength.

Reinforced concrete roofs can be designed to span in two directions but two-way action only becomes important in reducing the depth of concrete needed if the roof sections between supports are approximately square. When a roof section is twice as long as it is wide, two-way action is almost non-existent. Economical spans for reinforced concrete will be less than for the precast concrete. With 1½ - 2 feet of earth cover, spans should be kept below 20 feet and preferably to about 15 feet maximum. A slab and beam system will enable longer spans to be used than with a plain slab. The structural depth will, however, be increased.

Heavy timber roofs again span only in one direction with the planking merely providing a lateral transfer of load to the beams. Economical spans will depend on the size of beams available but should be similar to those for reinforced concrete.

A problem to be considered with all horizontal roof construction is that of ponding. Ponding is the tendency for a flat roof which has deflected under load to pond water causing increased deflections and hence further ponding in a cycle which can lead to failure in an otherwise sound roof. The critical parameter is whether the induced deflection will cause significant extra load from ponding. The problem is most serious for long span lightly loaded roofs. For earth sheltered roofs the progressive collapse potential should not be great because of the heavy permanent loadings. However, a camber should be built into the

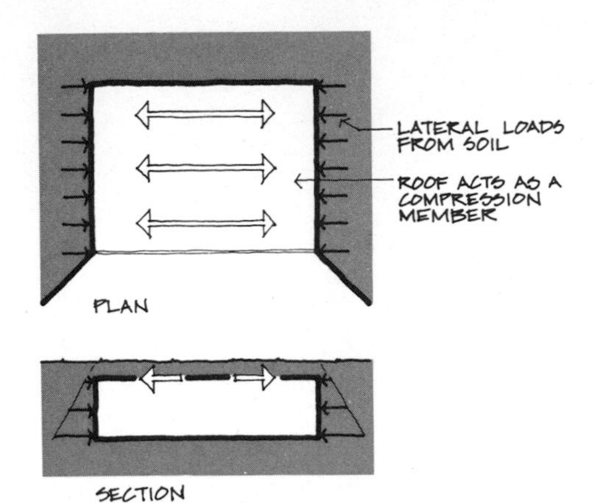

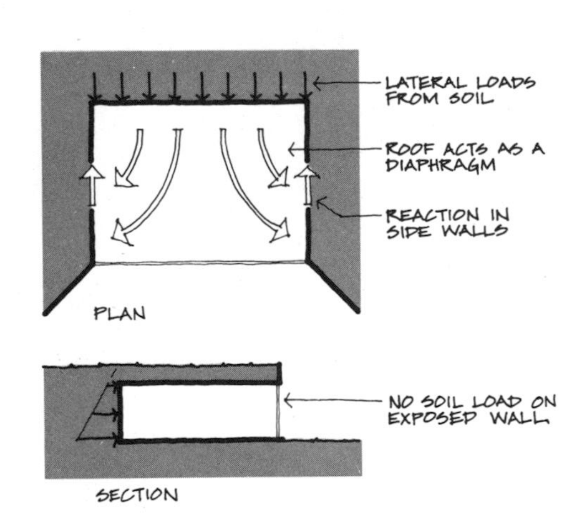

4-17 roof diaphragm action

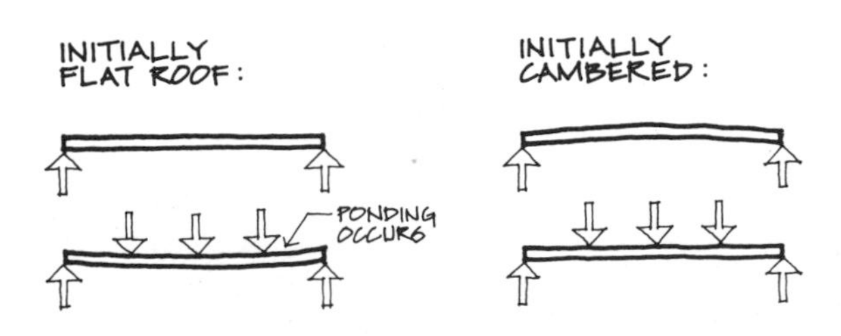

4-18 effect of roof camber

structure to avoid a downward deflection under self weight and the soil load. This will help avoid waterproofing problems that could be caused by slight ponding. The roof should also be designed for the saturated weight of soil rather than the dry weight. It should be noted while mentioning roof deflection that internal finishes are best applied after the roof has been covered since the earth load will deflect the roof structure slightly which could cause cracking in a brittle finish.

A further function of the roof will usually be to provide support to the walls of the structure which are retaining an earth pressure. This will involve compressive or diaphragm forces in the roof structure which must be considered in the roof design. For the reinforced cast-in-place concrete and the precast concrete roof this should not have great design implications for the roof. For the timber roof, sufficient diaphragm action and compressive strength perpendicular to the beams are hard to incorporate when the lateral earth loads are high. This is discussed further in the intermediate floor discussion.

intermediate floors

As mentioned under walls, intermediate floors in a two level design must provide the greatest reaction to the wall. The implications for the floor design are hence to avoid large openings in the floor adjacent to an exterior building wall resisting earth load. Sufficient floor width should be provided to enable that section of the floor to support the wall as a beam on its side.

Beacuse the wall reactions are so high on the intermediate floor, typical wood floor construction is not suitable when the floor must act as a diaphragm unless very frequent shear walls are used to limit the shear build-up.

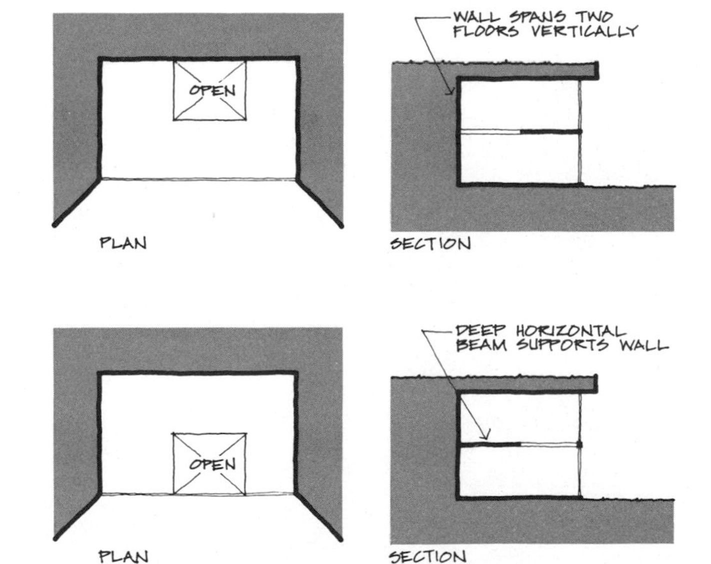

4-20 openings in intermediate floors

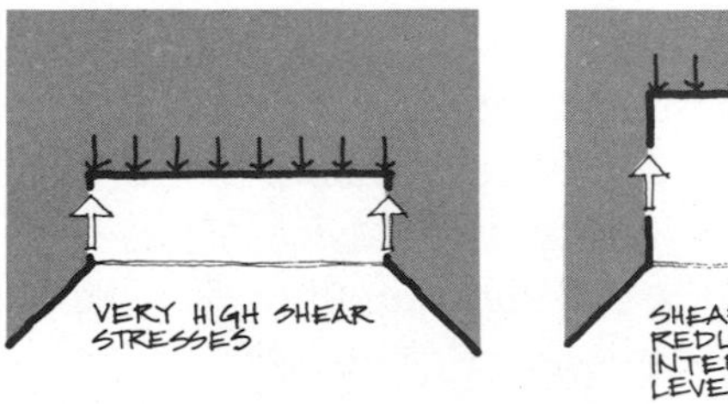

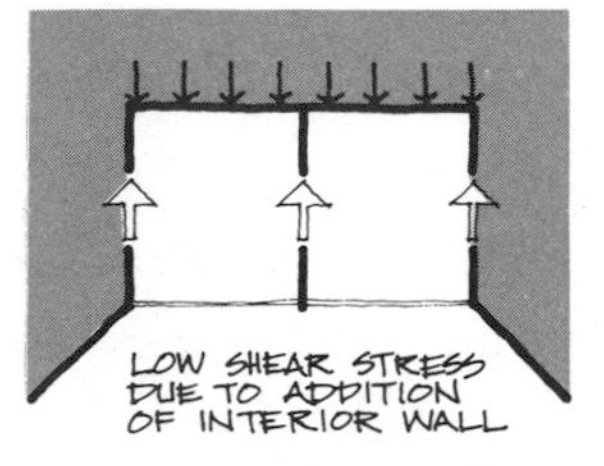

4-21 effect of shape on shear

interior building walls

These are only major load transfer components in a house when they are designed to act as intermediate roof or wall supports. The most common technique is to replace a non-load bearing stud wall partition with a concrete block wall, or in cast-in-place concrete construction, perhaps another poured concrete wall. Since there are usually openings required in these internal walls, steel beams are used to bridge the load across the opening if this is necessary.

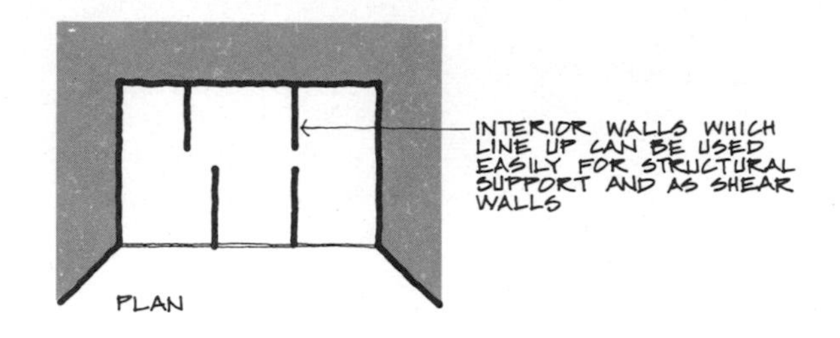

4-22 interior walls

ground floor

Generally, the ground floor of an earth sheltered structure is limited to a poured concrete slab since it is most economical and is the most suitable overall.

Unless under floor pressures are expected to be present from sources such as water pressure or squeezing soil, the ground floor is poured at a nominal thickness with 3½ - 4 inches generally being a minimum. The ground floor does not usually have to act as a diaphragm when it supports unbalanced wall loads. Instead, the forces are distributed directly to the ground in shear between the slab and the soil below. When two story wall construction is used the ground floor slab should be checked for adequacy in transmitting the necessary compression and shear.

Slabs subjected to underfloor water pressures will be much thicker and the floor will be transferring loads to walls and interior supports in much the same way as the roof. When the design water pressures are substantial, the conventional footing and slab construction is eliminated and a more continuous construction used.

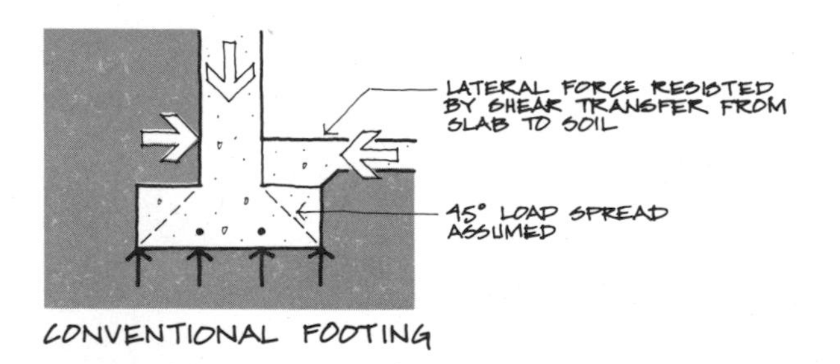

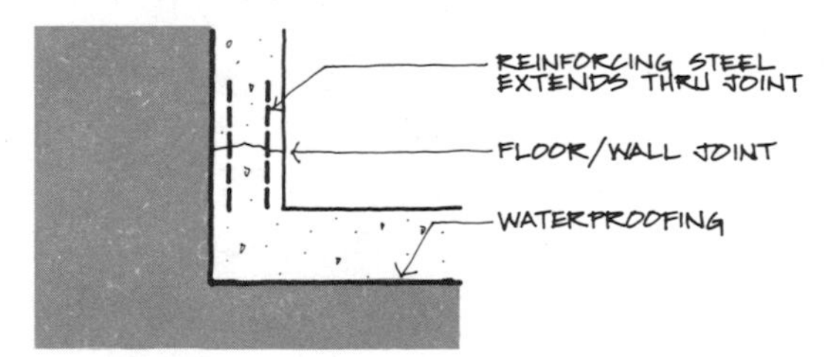

footings

Walls carrying the earth load from the roof may need larger footings than are customary for conventional basement walls. Code permissible bearing values given by general soil classification are usually very low. Advice on permissible bearing values should be sought as part of the soil investigation. When footings become substantially wider than the wall they support, they are usually reinforced rather than being deepened. They are assumed in design (without applied moments) to have an even bearing on the soil.

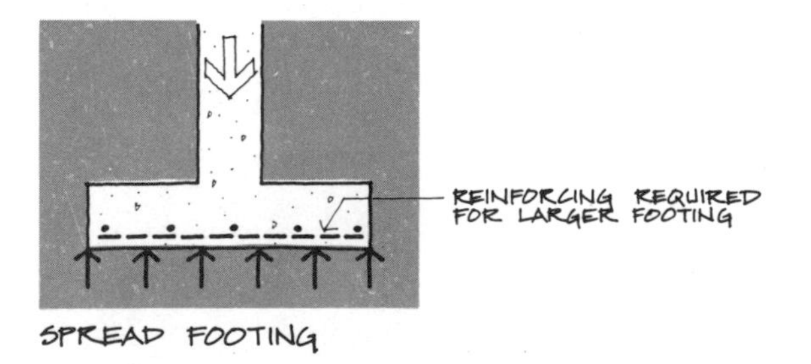

4-23 footings

outside retaining walls

These differ from the exterior building walls in that no floors or intersecting walls are available to provide useful support to the wall. Some of the techniques for retaining earth are illustrated and discussed briefly below:

Gravity wall - The mass of wall resists overturning and sliding.

Cantilever wall - Earth pressures are resisted by bending moments in the wall. The placement of the footing relative to the wall affects the stability of the wall against overturning and sliding as well as the amount of extra excavation required.

Tie Backs - These are generally protected steel cables connected to ground anchors in the soil. The wall (or facing as it really becomes) is supported back into the soil. The block of soil outlined by the dotted line then must be stable from its own mass. Tie backs can be used with the exterior walls of the building but would be unlikely to be used when an intermediate floor can limit the vertical wall span to 8 feet.

Reinforced Earth - Reinforcing strips are laid in the soil at frequent spacings. They resist any tensile forces that tend to develop near the face of the wall. The soil transfers the load to the reinforcement in shear and the reinforcement will be in tension. The facing is non-structural, providing resistance to erosion. Reinforced earth is most effective for high retaining walls when high shear stresses can be transferred because of the vertical pressure.

Cribbing - Cribbing can be carried out with concrete shapes railroad ties, wire mesh baskets, rubber tires etc. The earth is contained to form a steep but stable embankment. Cribbed retaining walls usually have a slightly backward slope when the height is substantial. Stepped retaining walls can provide an excellent opportunity for landscape planting to soften the appearance of the wall.

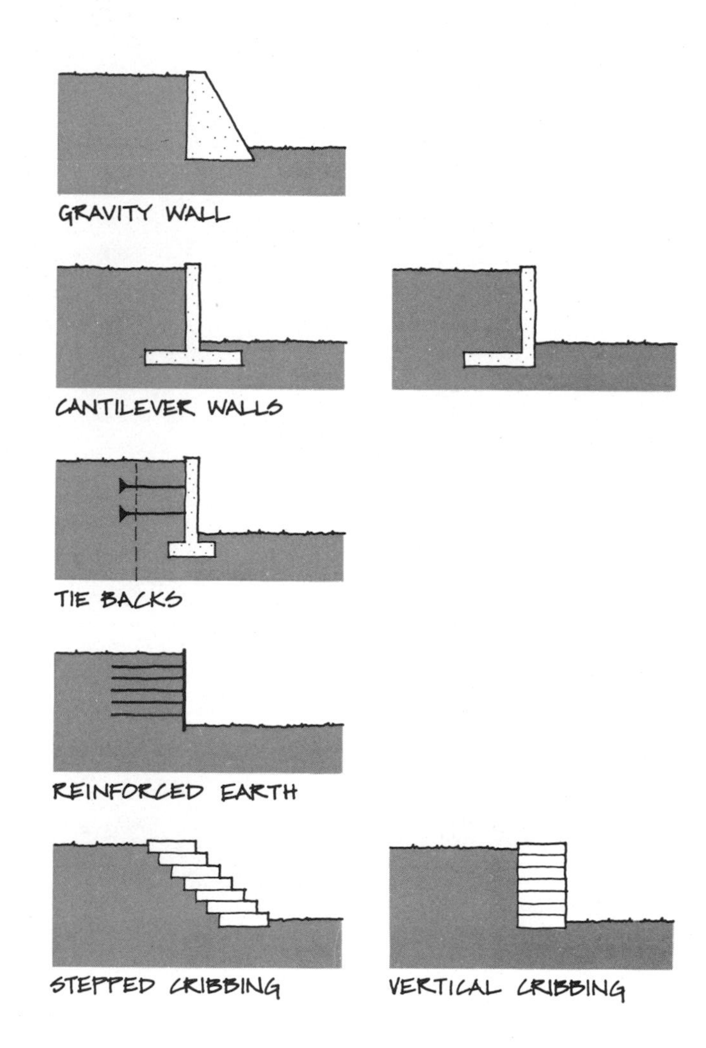

4-24 retaining walls

shell structures

Since earth sheltered housing has greater than normal loads on the roof and walls due to the earth cover, the use of curved shell structures to form the basic enclosure of a house should be considered. The geometry of a curving structure acts to support the loads without requiring as much material as a conventional flat roof system. This results in a more economical structure which can have the capacity to support very substantial earth loads. Shell structures of steel or concrete may become costly for a single home since more complex design and construction methods may be required for these unconventional systems. The costs associated with custom design and construction of a shell structure may be reduced or eliminated if the components are readily available. One such system is discussed below and an example of a steel shell structure for an earth sheltered house appears in Part B of the report.

Corrugated steel plate culvert sections are one type of arch or shell structure that is readily available. Curved corrugated plate sections are bolted together to form the complete structure and spans are available up to 40 feet. Generally, these structures can easily withstand several feet of earth cover; in fact, they may not be stable if sufficient earth cover is not provided. Thin wall structures like these are most effective when uniform pressures are exerted on the arch. Interruptions in the continuous profile generally introduce local bending stresses and should be avoided if possible. Two major concerns should be addressed when these structures are used for a dwelling:

1. The plate arches are generally used for drainage and highway structures where the structure can easily be inspected from the inside and the structure can be relatively easily replaced if problems occur. This will not be the case for a dwelling structure and design safety factors should reflect the greater consequences of a poor performance.

2. Corrosion is always a problem with steel structures underground and again, because of the use for which the construction is intended, greater attention should be paid to ensuring a long life structure than is necessary for the normal uses of the plate arches.

   Further information on the design of steel plate arches can be obtained from suppliers (see reference section).

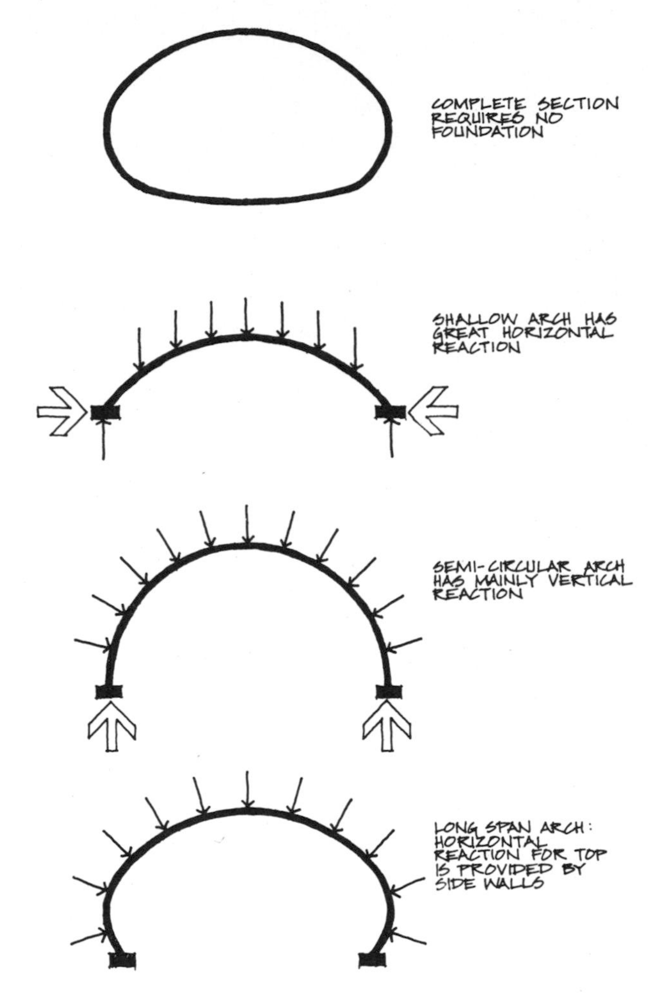

4-25 shell shapes

Three basic house configurations are examined briefly to point out some of the structural options available and the critical elements for design.

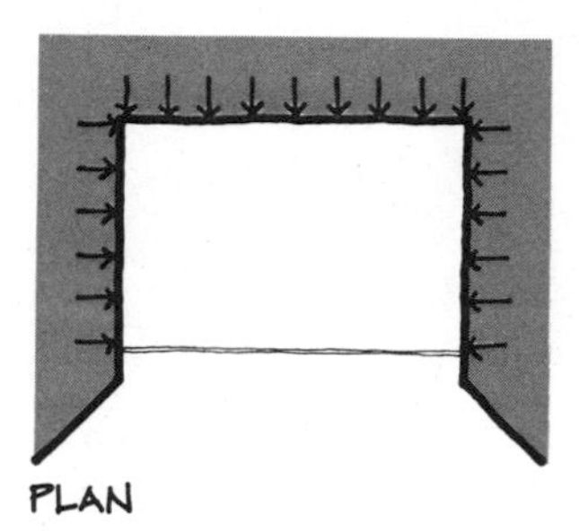

4-26 elevational layout

elevational — one level

- roof must act as a diaphragm
- roof can be flat or sloped
- roof can span from back to front acting as a cantilever for an overhang. This is difficult to insulate properly, however.
- roof can span from back to front with a separate structure for an overhang.
- roof can span in the long direction with intermediate beams or bearing walls. Cantilever can be a plank separated by a thermal break from the rest of the structure.
- Rear and side walls span vertically between floor and roof retaining earth pressure.
- side walls also act as shear walls to transfer roof diaphragm loads to the ground.
- wing retaining walls are needed to allow complete burial of the side walls. A thermal break should again be included bewteen the building walls and any separate structure which is a relatively poor insulator (such as concrete).
- critical design parameters will probably be earth load on roof vs. span of roof and the design of the vertical span for the retaining walls. Diaphragm stresses may be critical in wood construction.

elevational — two level

- the comments for a one level design still apply, the intermediate floor must carry the greatest diaphragm load.
- critical design parameters will be the roof load vs. span, the design of the two story retaining wall and the diaphragm capability of the intermediate floor.
- end shear walls should be kept relatively free of openings. An intermediate shear wall may be necessary for a long narrow structure.

atrium — one level

- the earth loads are more in balance from one side of the structure to the other than in the elevational scheme. Diaphragm action is needed opposite the atrium opening to transfer the force from the wall to the remainder of the roof slab.
- diaphragms will tend to be narrow and hence should be kept fairly short unless intermediate shear walls are used.
- internal bearing walls or beams will be needed to provide support for the roof slab at the corners.
- shear wall action will tend to be minor, being confined to the end walls adjacent to a U-shaped atrium.
- critical design elements will be roof load vs. span, retaining wall design and placement of interior supports for the roof.

atrium — two level

- two level atriums would generally require a larger atrium size to admit sufficient light to the lower level.
- the larger court size coupled with the high earth loads will give much higher diaphragm stresses than for the one level.

penetrational — one level

- the earth loads are balanced across the structure and hence the roof will be in compression from the wall loads.
- all walls can span vertically between roof and floor.
- the retaining walls at the openings are best designed as separate structures to allow thermal breaks to be installed.

penetrational — two level

- if a smaller second story is used in a two level design, it will be difficult to completely cover the structure because the earth loads will become excessive on the roof of the one story section.

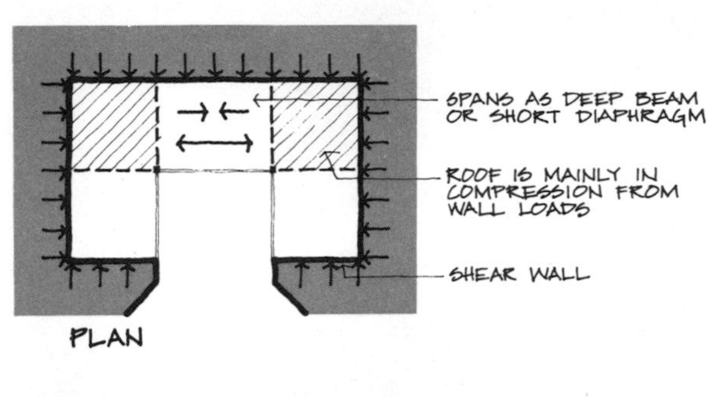

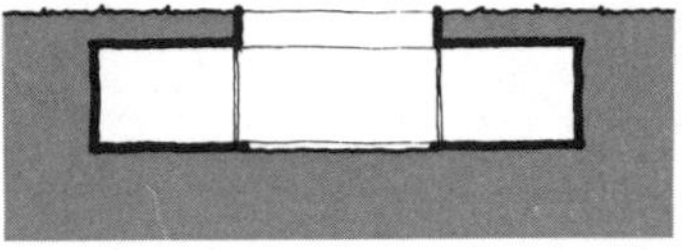

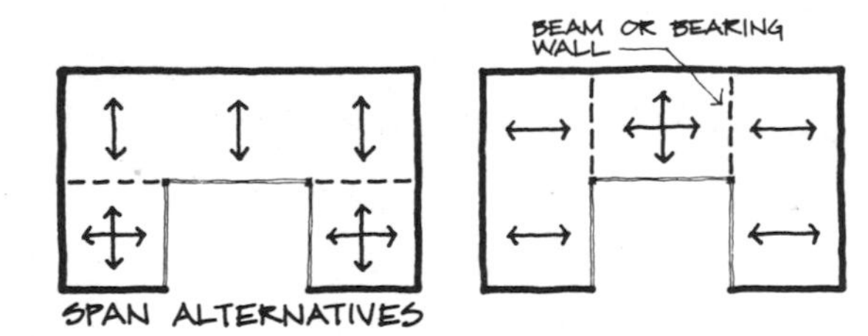

4-27 atrium layouts

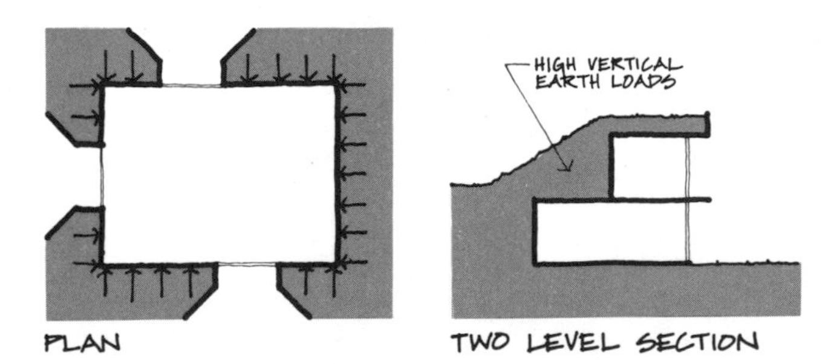

4-28 penetrational layouts

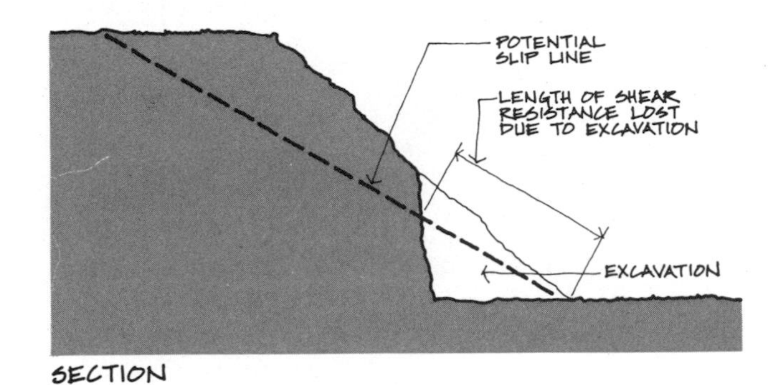

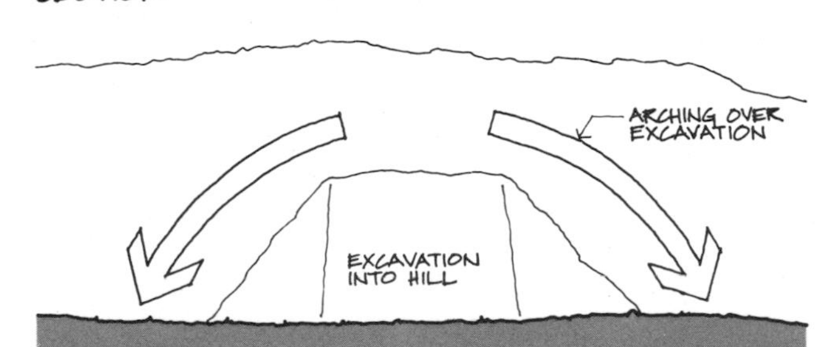

4-29 slope stability

slope stability

Sloping sites are often an ideal situation for an earth sheltered house but when designing with a steep slope, the overall stability of the slope must be considered as well as the normal earth pressures to be retained. Steep slopes, although stable in their natural state, may be close to a critical slope angle. Cutting into the slope for house construction will weaken the slope unless special precautions are taken. In a long slope, this problem may not become critical until several adjacent excavations have been made. This is because the slope will bridge or arch around a small excavation and the effect of the one excavation on the whole slope may not be critical. As more and more excavations take place, the effect on the slopes becomes larger.

backfilling excavations

Care should always be taken when backfilling excavations to ensure that suitable materials are used (see drainage section), and that the backfill is properly compacted. Lumps of clay or frozen soil will not compact well when the backfill is first placed and later the backfill will settle causing a low area adjacent to the house. This will be especially noticeable in backfilling two story retaining walls. The backfill should be compacted in layers and these layers are generally specified to be no more than 9 - 12 inches thick for normal civil engineering projects where the backfill quality and density are important. Backfilling in layers of this thickness requires less force for compaction than when thick layers are used. The forces exerted on the wall can be greater during backfilling and compaction than when the wall is complete. Temporary shoring of the wall can be used but unless preloaded against the wall (using wedges for instance) the deflection of the support before taking significant load may make them more cosmetic than useful. Well graded granular backfills (e.g. coarse sands and gravels) will require little compaction and are the easiest backfill materials to handle.

stepped retaining walls

Stepped retaining walls can be used to reduce the structural member sizes required to resist earth pressures. In the adjacent diagram the walls of the structure are stepped so that each retaining wall is founded on a stable slope angle for the soil. In this condition, the soil beneath the dotted line is stable and need not be supported. The top retaining wall thus supports a wedge of soil bounded by the ground surface, the wall and the stable slope angle. The lower retaining similarly supports the lower wedge of soil except that any loads from the upper floor on the soil will add to the earth pressure at this level. The effect on the design earth pressures can be seen by comparing the stepped wall pressures with those for a straight wall.

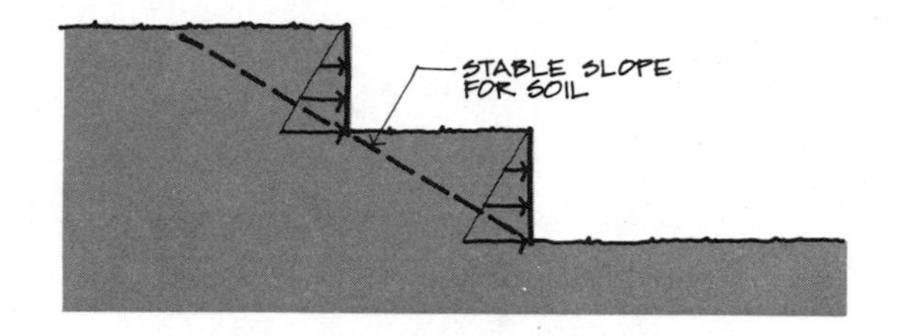

4-30 stepped retaining walls

5 waterproofing and insulation

- sources of moisture problems
- drainage techniques
- damp-proofing methods
- waterproofing methods
- costs of waterproofing
- insulation

introduction

The possibility of a moisture problem is one of the biggest concerns of people considering an earth sheltered home. The generally poor performance of normal basement construction with respect to moisture problems seems to be the major cause of concern. It must be remembered, however, that basements were not considered as living space until recently. Basements were originally provided because the footings for a house had to extend below the frost line and, in the northern parts of the country, this entailed substantial excavation. It was realized that useable extra space could be created for very little additional cost if a basement rather than just a buried foundation was constucted. Because the basement or cellar was simply a cheap addition of storage and utility space, little or no waterproofing measures were taken.

This attitude towards waterproofing basements has continued to the present day even though living space is now often anticipated in a basement. Some damp-proofing may be applied to basement walls, but as will be discussed later, the effectiveness of conventional damp-proofing is very limited. Forming an opinion in this way, is like looking at a barn and deciding houses shouldn't be built of wood above ground because the wind will blow right through. The requirements of construction between a barn and a house are very different and hence so are the standards of construction. The same is true for basement storage space compared to an earth sheltered home.

There is no denying that moisture problems can be troublesome and that care in the selection of a waterproofing system, and more especially, care in the application is needed. If this care is taken, however, there should be no need to fear a damp, leaky underground home.

The cost of proper waterproofing is not negligible but it can be offset in a comparison with a conventional house against the cost of an attractive and weatherproof siding. For example, the cost of aluminum or redwood siding which both have good permanence and low maintenance is approximately $l.75 per square foot including backing felt (and stain for the redwood siding), but not including sheathing or insulation. The cost of full waterproofing of a below grade structure will range from about $0.50 — $l.50 per square foot depending on the technique used. A perfectly acceptable solution will be available for most conditions in the $0.70 — $0.90 range.

surface run-off

The first steps in waterproofing a house should be taken when the site is selected and preliminary layouts and landscaping are considered. At this time, the main aim is to avoid as many water problems as possible rather than have to design to cope with them later.

The first decision is obviously connected with the choice of site. A perfect site from the waterproofing aspect will rarely be found and is not necessary, but a good site will save money and potential problems as opposed to a bad site. Specific points to be looked for are listed below with diagrams of the problems and potential solutions shown where appropriate.

Low areas and flood plains

These should be avoided with any type of housing if possible. The potential danger to an earth sheltered house compared to a conventional house depends very much on its layout or design type. An earth sheltered house of the berm type built above grade with earth bermed up around it will not be in any more danger from flooding than a conventional house and would be far less likely to receive serious structural damage. An atrium design located completely below grade, on the other hand, (such as the Ecology House shown in Part B) has a much greater potential danger in a low lying area. Such structures are not naturally self draining and drain pipes and possibly sump pumps must be installed to handle any water which collects in the court area. These drains and/or pumps are sized to handle expected rainfalls or ground water flows but would not be capable of handling a large flow of surface water. Most other designs will have intermediate implications between these two extremes.

If a low lying area is being considered as a potential site, the contours of the surrounding area should be studied closely to determine the low spots where water will tend to collect during heavy rains or snow melt.

Gullies

An allied question to the avoidance of low areas is to avoid sites, even on higher ground, which are potential gullies for surface run-off during heavy rains.

SECTION A

SECTION B

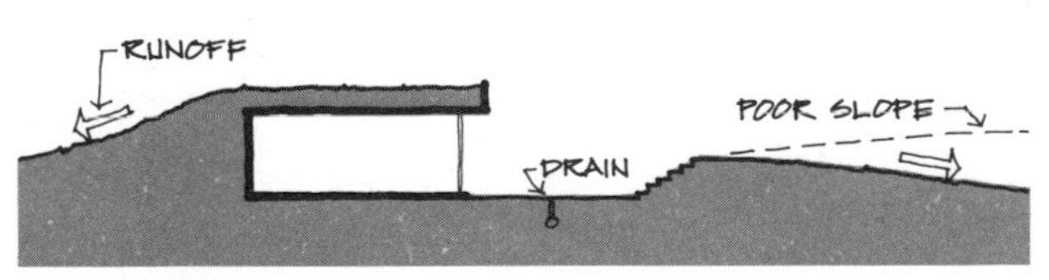

SECTION C

5-1 drainage techniques

Ground slopes

Having identified the possibility of potentially disastrous surface run-offs, more gentle run-offs and water percolation must be considered. Here again different designs and different sites have different implications as shown in the adjacent illustrations.

The first two drawings show different methods by which water can be diverted away from an earth sheltered structure at the base of a slope. Section A illustrates a drainage swale or gully which diverts the run-off around the house, while Section B shows a cut-off gravel trench with drain tile at the base. Section C illustrates some drainage problems associated with sunken courtyards which are common in earth sheltered designs. It is preferable for the surrounding ground to slope away from the structure on all sides, then only the rainfall which falls directly into the court must be handled by a drainage system.

The grading of the roof area will depend on the waterproofing technique used, the type of soil used on the roof and the water requirements of the vegetation on the roof. It is not a good idea to pond water on the roof even if a high quality waterproofing is used, but too rapid a drainage would leave the earth with too little water soaking into support the vegetation. A slope of between l percent and 5 percent will provide such gentle drainage, but as can be seen from the examples in Part B much steeper slopes can be used successfully. With steeper slopes the type of soil used must be able to hold water in order to easily support vegetation.

water table

A high water table, like a low lying site, preferably should be avoided. If a site with a high water table is selected for other overriding considerations, however, this problem can be overcome in most situations. One technique is simply the use of a drainage svstem to draw down the water table around the structure. This is discussed in greater detail in the following section on drainage techniques. If high water exists in very permeable material, a drainage system may not be able to drain the area fast enough and a substantial waterproofing system must be applied to the entire structure. Often the drainage system will be used in combination with the waterproofing. Various waterproofing techniques are discussed in a later section.

temporary water pressures on walls

Temporary water pressures can occur even when the foundation of a house is substantially above the water table. Seepage of water into the ground adjacent to the house faster than it can soak away below to the natural ground water table can cause this during heavy rainfalls. This is mitigated by sloping the ground away from the house and making sure that the backfill is properly compacted so that it does not later slump to trap water and so that large voids are not left which can rapidly fill up with water during a rainfall. Where free draining backfills are used, these should be capped with a properly compacted layer of natural ground or clay or other impermeable material to stop the free draining material from acting as a catch basin for water from the roof. In connection with this, gutter systems around an exposed roof together with the horizontal leads that take the water from the roof well away from the house will prevent an additional water load close to the underground walls.

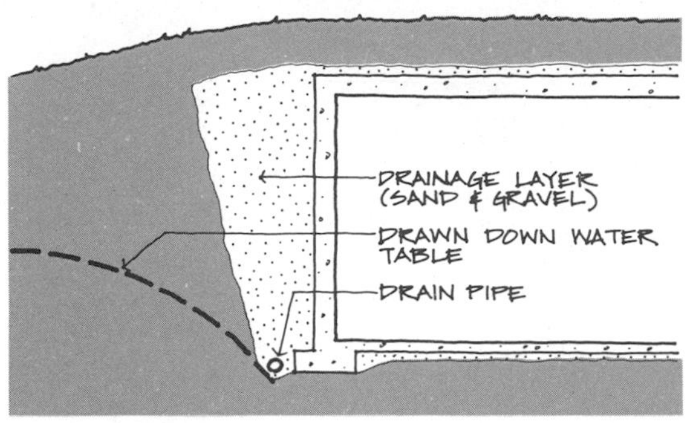

DRAINAGE

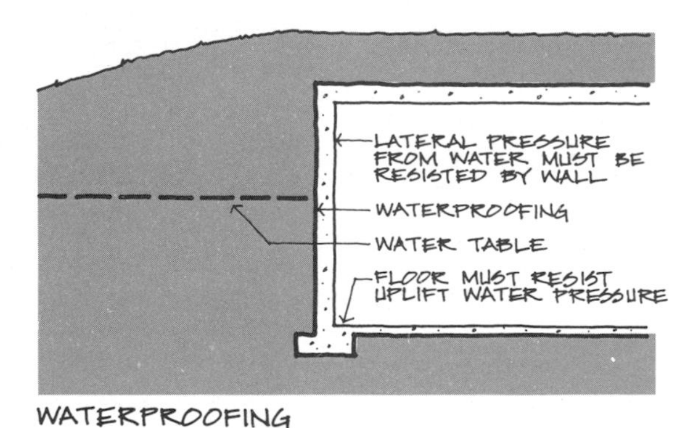

WATERPROOFING

5-2 water pressure on walls

Temporary water pressures can also occur in the spring because the ground closest to the house will usually melt first due to the heat loss from the house. The ground further away from the house will still be frozen and impermeable. Again, a natural sump adjacent to the house will be created and water from melting snow or early spring rains will tend to build up in this area and apply a positive water pressure to the walls. The natural tendency for there to be a crack or high permeability plane between the backfill and the foundation wall increases this problem.

This build-up of temporary water pressures is the biggest complication in designing a moisture control system for a foundation above the normal water table. If temporary water pressures could be eliminated, only a damp-proofing technique would be needed to stop any capillary draw of moisture into the wall. When temporary pressures cannot be eliminated, and this could rarely be guaranteed, damp-proofing will not suffice for complete protection because damp-proofing is not designed to keep out water under pressure. This fact is amply demonstrated by many houses which may or may not have damp-proofing and are very acceptably free from damp almost all the time but can periodically leak in the spring or after very heavy rains. An occasional slight leakage may be tolerable in storage or utility space but will not be tolerable in a prime living area. It must also be stated, however, that many houses are constructed with little or no damp-proofing and never have any water problems. The question for the home builder to decide is whether he is prepared to take a risk of water leakage in order to save the cost of full waterproofing.

vapor transmission

It is often thought that water vapor transmission through the wall is the cause of moisture from the ground making unprotected underground spaces damp. This is not, in fact, the mechanism by which dampness enters such a space. The pressure which water vapor in the air exerts depends on the temperature of the vapor — the higher the temperature, the higher the vapor pressure. Since gases or vapors always tend to move from an area of high pressure to an area of low pressure, and since the temperature inside an earth sheltered house will almost always be warmer than the ground around it, the water vapor transmission will tend to be from the inside of the house towards the ground. In a conventional house, a vapor barrier is placed on the warm side of the construction to prevent the water vapor from reaching a cold part of the wall where it could condense and could cause damage to the structure. The same is true for an earth sheltered house. If the vapor reached a part of the structure which is below the dew point for the amount of water vapor in the air, condensation will take place. This is discussed further in the consideration of insulation placement. Hence, it does not really matter whether a waterproofing material is a good vapor barrier in addition. This property does often go hand in hand, however, with the requirements for the prevention of capillary draw discussed below.

capillary draw

This mechanism is responsible for most of the dampness problems (other than condensation) which occur in basements. Moisture from damp earth can be drawn into the wall by capillary suction. This is the same mechanism by which a sponge laid on a wet surface will draw water up into itself. This moisture will be drawn through the wall and if the air inside the basement is not saturated, the moisture will evaporate and raise the relative humidity. Whether dampness appears on the wall or not will depend on how fast the moisture can be drawn into the wall relative to how fast it can evaporate. If evaporation is slow because the inside air is humid or cool and substantial amounts of moisture are being drawn into the wall from outside, the surface will become damp. If the evaporation rate is high, the moisture may evaporate before it reaches the surface of the wall. The wall will then appear dry but a substantial increase in the humidity of the basement air could still be occuring.

There are two methods of breaking this capillary draw. The first is to cut it off with an impermeable barrier such as a waterproofing or damp-proofing application. This is the use for which damp-proofing products are mainly intended. They can interrupt the capillary draw and small defects in the barrier are not as important for this application as when trying to stop water flow under a positive pressure. A more effective means of capillary break is the use of an air gap or a material that is so open that moisture will not be drawn through it. Swedish and Norwegian research (see bibliography) has indicated the effectiveness of using either an air gap or a 2 inch layer of rigid mineral fiber insulation in keeping basement walls dry. In fact, the use of the open weave insulation in preventing moisture problems illustrates the fact that it is not necessary to have a vapor proof layer on the outside of the wall as long as the capillary draw is interrupted. It must be repeated here, however, that the open capillary break system will not provide any protection once more water in the ground is present than can be drained down the outside of the capillary break. If more water than this builds up, the air space or insulation will fill with water and will provide no protection at all. The basement wall designs resulting from the Swedish research allow for some build up of water pressure by providing a 20 inch freeboard above the foundation using a membrane or double asphalt coat protection.

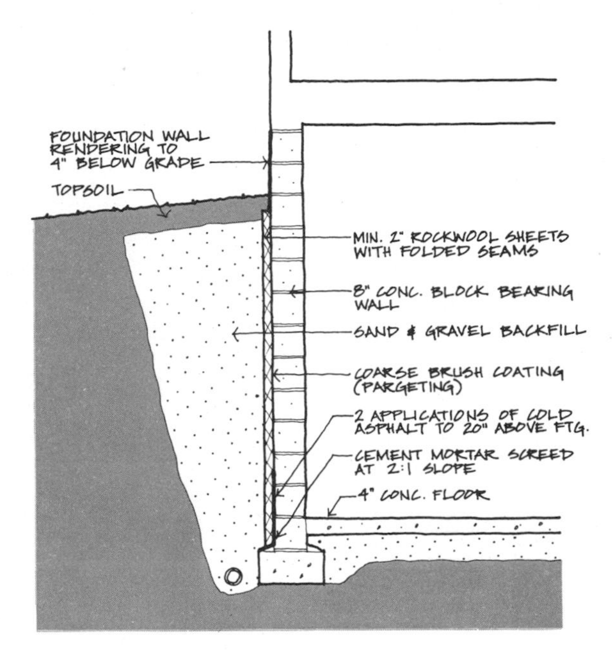

5-3 swedish system

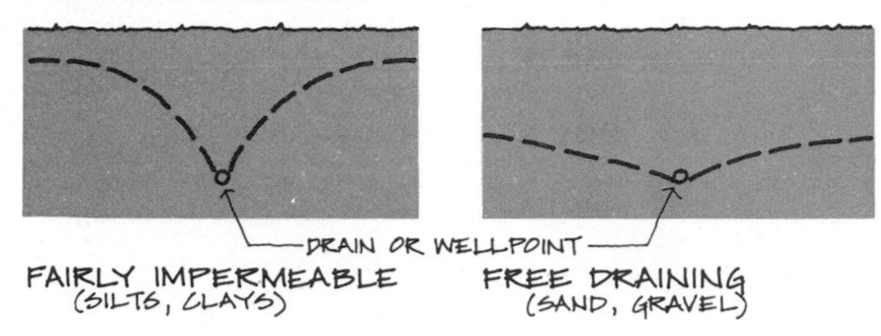

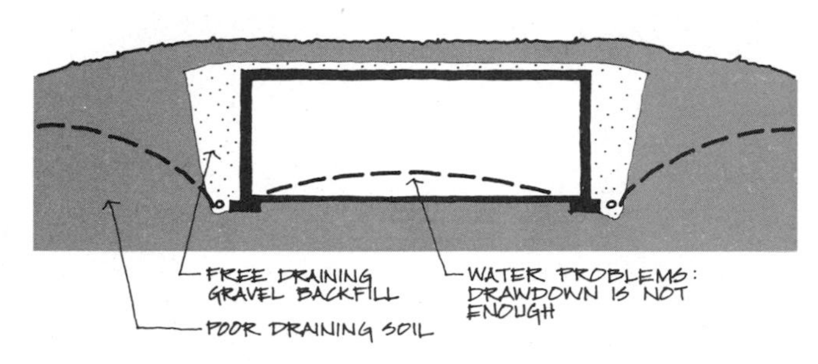

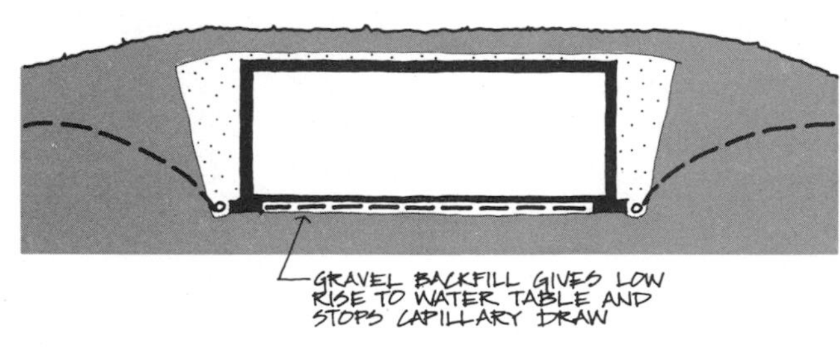

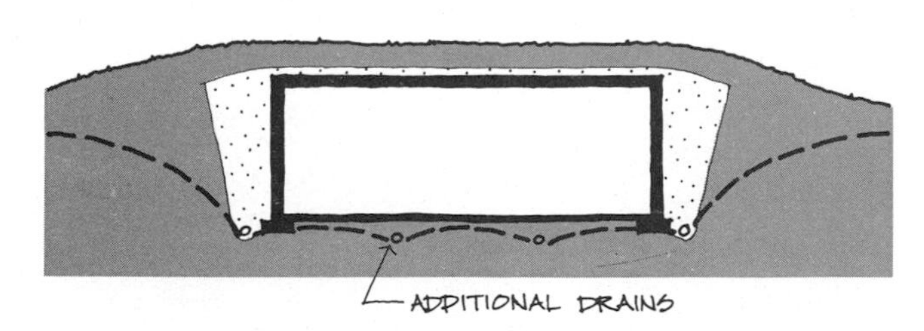

5-4 drainage techniques

Drainage techniques should always be an important consideration in a waterproofing system since they reduce the frequency and duration of water conditions that actually test the membrane. The use of drains to lower the ground water table will also eliminate the necessity of structurally designing the walls and floor for positive water pressures. It is important to locate the drains as low as possible with respect to the foundation and the finished floor level especially if the drains are being used to permanently lower a water table. Even if the drains are functioning perfectly, the water table will only be held to the level of the drains at the drain itself, the water table will rise away from the drains at a slope which is dependent on the permeability of the soil.

A number of points about the design of a drainage system are listed below.

1. The backfill should preferably be of a free draining material except for a capping of soil of low permeability.
2. The backfill should act as a filter material preventing the washing of clay and silt particles from the surrounding ground through the backfill and into the drain. This can cause settlement problems and also clog the drain. To further prevent clogging of the drain, purpose designed filter fabrics are available which can be wrapped around the drain. The necessary particle sizes and gradations for the backfill to act as a filter and yet provide sufficient drainage could be discussed with the soils firm which does the site investigation.
3. If the drawdown curve of the water table will not be sufficiently shallow to keep the water table below the floor level with only perimeter drains, additional drains should be placed under the floor.
4. The drains will preferably be able to daylight naturally on a sloping site. On a flat site they can be drained to a storm sewer or to a sump where the water can be pumped out whenever sufficient amount has collected. Drains cannot normally be discharged into a sanitary or combined sanitary and storm sewer system so where daylighting is impossible, the water must be pumped up and either allowed to flow away or be sprinklered out over the ground surface.
5. It is very desirable that some access for cleaning out the foundation drains should be incorporated into the design.

damp-proofing methods

Damp-proofing products are those which will interrupt the capillary draw of moisture into the walls of a building but are either too thin, have too little resistance to bridging wall cracks or deteriorate too easily to be considered as a good waterproofing technique. Although damp-proofing can help dampness problems in many instances, there are no guarantees with it. It is not intended to stop water penetration and hence there is almost no comeback for a system that doesn't work. Some of the techniques are discussed below.

concrete mix design and additives

There is much that can be done to make concrete more impermeable than it normally is, including keeping the water/cement ratio as low as possible, using adequate vibration when placing the concrete and including well distributed reinforcement to limit the size of the shrinkage cracks that occur when the concrete drys out. Admixtures can fill the capillary channels in the concrete, others react when they come into contact with water. All these techniques will cut down the permeability of the wall, but as long as concrete cracks and shrinks when it dries, which is always the case, (unless very expensive expansive cements are used) cracks will occur which destroy the integrity of the seal. For limiting the flow through the wall which is the primary goal in reservoirs, for example, the techniques can be very successful but even very small amounts of leakage can make very unsightly marks on the wall, roof or floor finishes.

pargeting

This refers to the coating of the wall with a dense cement plaster or similar material. This can be done on the inside or the outside of the wall, but as with all waterproofing applications it is best done on the outside of the wall so that any water pressure will not tend to force the coating off the wall and so that the wall will be on the dry side of the waterproofing (particularly important when reinforcing steel is contained in the wall because if it were on the wet side of the waterproofing it could then rust and spall the concrete). Cracking of the wall which almost inevitably occurs will also crack these generally brittle materials, and hence, they cannot be relied on as waterproofing materials.

asphalt coating

These can be applied hot or cold and can be sprayed, brushed or trowelled on. In general, trowelling gives a better job than spraying because a denser layer with better adhesion is obtained. Hot asphalt applications become brittle when they cool and the cold asphalt applications have some bridging characteristics over cracks but not sufficient to bridge normal sized structural cracks. Asphalt emulsions are slowly soluble in water and the quality of asphalt supplied to the building trade has deteriorated over the past few years.

pitch

This is a more stable material to use underground than an asphalt emulsion. A special waterproofing pitch used to be available but was discontinued to comply with health standards of fume release during application. The major pitch available now is a general purpose pitch designed for roofing applications. It has a softening point of about 150 °F and hence underground it will remain brittle with no resealing abilities.

polyethylene sheet

This a very cheap material to use. It degrades when exposed to sunlight but when completely covered underground it should last a long time. With reasonable laps and care taken not to puncture the sheet during placement and backfilling it will function as a good barrier against capillary draw and vapor transmission. No attempt is usually made to seal the laps completely and hence it will not resist a water pressure on its own. It does have the capability of bridging some cracking in the concrete. For the floor of an earth sheltered house that is above the water table, has a foundation drain system and gravel layer under the floor, this type of damp-proofing protection will usually be adequate.

liquid seals

These suffer from the same problems as most damp-proofing techniques. When the wall cracks the integrity of the seal goes with it.

waterproofing methods

Techniques other than those intended for dampproofing are considered here. Not all of the techniques listed are suitable for waterproofing against a continuous water pressure and this will be noted in the discussion. The list is clearly not all inclusive and any products or techniques not listed here should be examined with the advantages and drawbacks of the various listed systems in mind.

built up membranes

These consist of layers of asphalt or pitch alternated with felt or fabric reinforcing. Three to four plies would be considered a minimum for waterproofing purposes. Built up membranes can be used on the roof or walls of the structure. The same remarks about asphalt and pitch made during the discussion of their use as damp-proofing products apply to their use in membranes. The fabric does give some mechanical strength but little elasticity. Glass fiber fabric should be used rather than organic felts because the felts will rot if they become exposed to water and replacement is not as easy as with a conventional roof. There was a difference of opinion between the various waterproofing specialists contacted during the study as to whether a built up membrane offered sufficient waterproofing protection. The big disadvantage of any membrane like this is that if the membrane does leak in one place the adhesion is usually not good enough to prevent water from travelling behind the membrane and appearing inside at a point remote from the actual leak. This makes finding leaks a difficult and expensive business especially for an underground structure. On a vertical surface it is difficult to apply pitch at a hot enough temperature to saturate the glass fabric. All membranes are susceptible to damage between the time they are applied and inspected and the time after they have been backfilled. Membranes must be elastic enough to bridge cracks in the structure or they must have a resealing ability. A built up roof with asphalt or pitch does not have very good bridging characteristics, nor does it have any resealing capability at undergound temperatures. Nevertheless, it has been used in a number of earth sheltered houses and large underground structures with success and was the choice of one of the waterproofing specialists contacted for an earth sheltered roof.

bituthene

This is a polyethylene coated rubberized asphalt. It has sufficient adhesion to allow it to be rolled out and applied to a wall or roof surface. This membrane must be buried or covered to prevent ultra-violet induced deterioration of the polyethylene but once covered it should have a long life. Its bridging characteristics are reasonable. It is not intended for use under a continuous water pressure and it is recommended that when applied to roof surfaces the membrane should have a slight slope to give positive drainage for the roof. Poor adhesion will be obtained if the surface to which it is applied has moisture on it or is below about 45°F in temperature. General disadvantages of membranes apply.

polyethylene embedded in mastic

Here mastic would be applied to the wall of an earth sheltered structure and sheets of polyethylene with generous laps (also sealed with mastic) embedded into it. This would be unlikely to suffice for continuous immersion in water but may provide adequate protection against occasional water conditions. It is not recommended for the roof of an underground structure and should be used with a foundation drain system. General disadvantages of membranes apply.

butyl rubber, epdm, neoprene membranes

These materials are all quality materials and have good bridging characteristics. The major problems with these membranes is making the proper seams between the sheets of material under field conditions. This requires great care and close inspection. Even though the materials are fairly tough, accidental damage after installation and the water travelling behind the membrane are still problems. Compartmentalization (where the material is glued down with glue lines at regular intervals) is often used to reduce the travel of water from a leak. This enables the leak to be found more easily than would otherwise be the case. The material can also be completely bedded down but this adds substantially to the cost. The experience of people involved with waterproofing with these products is quite varied. Some has been very poor and others excellent. Malcolm Wells uses Butyl rubber and EPDM membranes on his underground structures and has had very good success with full bedding of the membrane and tight inspection of the process. These membranes can be used for the roof or walls of the structure. They should be water tested if at all possible.

liquid applied polymer membranes

These are usually polyurethanes and are preferably trowelled on. Liquid membranes do not need seams and can easily be used to seal in awkward areas or around vent pipes etc. There is, however, some loss of control over exactly how thick a membrane is applied and whether all parts of the surface are adequately covered. The materials generally have reasonable bridging characteristics but no resealing ability. In general, these products are used where conventional membranes would be awkward to use. Surfaces should be clean and dry for application. Liquid membranes would generally not be recommended for use on precast roof systems. Otherwise they can be used for roof and walls.

bentonite panels

These are cardboard panels filled with bentonite which is a clay which expands greatly when it comes into contact with water. As the clay tries to expand it seals itself against further penetration by water. It can be applied by unskilled labor without difficulty and can be nailed to vertical surfaces. The cardboard rapidly decomposes in contact with water so the panels must be protected from rain until the backfilling is complete. Because of its swelling characteristics,bentonite does have some resealing ability as well as a bridging capacity. There is some concern about the cardboard which remains behind the bentonite providing a lateral water transmission path. The material is natural and not subject to degredation with time and it has been used successfully for continuous water immersion. It should not be used where running water could slowly carry the bentonite away from the wall and it should not be used where there is a high concentration of salts in the ground water since this interferes with the swelling mechanism.

spray-on bentonite (bentonize system)

In this material,the bentonite is mixed with a small amount of mastic binder so that the material can be sprayed onto a wall or roof and will adhere to it. It must be protected from rainfall prior to backfill although it is more resistant to slight wetting than the panels. The material is sprayed on three-eights inch thick and will provide complete waterproofing for a structure with the bridging and resealing properties described above. Water does not easily travel behind the membrane and hence leaks are localized. A limited five year guarantee for any repairs necessary to the membrane is offered when the material is applied by a licensed operator. As with any spray-on membrane, the biggest problem with this material is making sure that complete coverage of the required thickness is achieved. The waterproofing does not require a perfectly smooth wall and the thickness applied can easily be inspected without damaging the membrane. Several houses in Minnesota have already used this system and bentonite in general was recommended by two of the waterproofing specialists contacted.

waterstops

These are of various types and are used to reduce water leakage at cold joints or expansion joints in concrete. Joints in poured concrete walls, floors etc. are places where most leakage occurs and the waterstop simply provides a much longer leakage path for the water. Water can tend to travel laterally along these waterstops and unless the joints between waterstop sections are perfect, leakage can occur at these points. For this reason, they are not recommended when a full waterproofing system is used on the outside of the wall. If water gets through the main membrane, the waterstop can mask where the leak in the membrane really is and makes leaks extremely difficult to find. Waterstops are used most when the reduction of the amount of water leakage is of major importance rather than the elimination of all leaks.

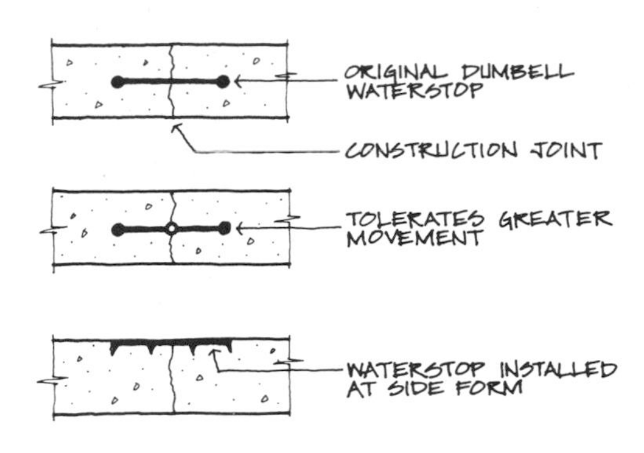

5-5 water stops

flashing details

One area of waterproofing which merits special attention is the treatment of corners and projections such as skylight wells or roof vents. These connection and termination points of the waterproofing system, generally referred to as flashing details, are very critical since they often represent the weak link in the system. With membrane sheet systems, generous laps should be provided whenever possible, and fillets are often used at corners to avoid sharp changes in direction. Non-sheet waterproofing systems are usually thickened at corners and joints. Since every product and every situation have different requirements, it is best to consult a professional in order to insure the proper technique for this important detail. Some examples of flashing details for various types of waterproofing appear in Part B of the report.

costs of waterproofing

The approximate cost of the various waterproofing or damp-proofing systems is shown below. These are estimates obtained from various sources and represent 1977 costs. Costs will vary according to size and complexity of job, location of job and whether the contractor is busy or not. The costs shown here should only be used as a guide in comparing different waterproofing systems. Local estimates should be obtained before any decisions are made. The costs include the application but not the protection of the membrane since protection can usually be provided by the insulation.

method	approximate cost/sq. ft.
dampproofing	
ironite pargeting	$0.75–1.10
asphalt or pitch coatings	
brushed on	.20–.35
sprayed on	.22–.25
trowelled on	.30–.45
addition coat-add	.20
polyethylene sheet–.010 in. thick	.13
.004 in. thick	.08
waterproofing	
built up roofing w/fabric	
1 ply	$0.35–.55
3 ply	.65–.90
5 ply	.85–1.35
bituthene	.75– .95
butyl, EPDM, neoprene membranes (without full adhesion)	1.15–1.50
liquid membranes	.70– .85
bentonite	
panels	.60–1.00
spray on	.50– .90

5-6 waterproofing costs

The function of any insulation is to reduce the unwanted transfer of heat from one area to another. Although this objective appears quite simple, it becomes somewhat complex when the insulation must perform within a harsh environment. Such is the case of insulation as utilized in underground housing. The specific thickness and locations of insulation have been discussed in other areas of this report. However, one fact should be restated: the most desirable position for insulation placement is outside the structure. This procedure allows the mass of the house to be located within the insulation envelope. Such mass prevents unwanted rapid temperature changes of the interior spaces; hence, HVAC equipment may be sized for average conditions rather than peak conditions and the internal environment remains at a comfortable constant level, rather than cycling. If the placement of insulation outside the structure is not possible, then it may be placed inside the structure, although this method is not as desirable from a comfort or space point of view.

Considering that placement of the insulation on the outside is advantageous, the following characteristics of insulation materials are desirable:

1. High compression strength to resist the lateral earth loads imposed by the backfill (20 to 30 psi is sufficent).
2. High resistance to water and very low water absorption so that the R value of the insulation is not reduced.
3. High resistance to the various chemical properties of soils; therefore, long lived.
4. Good dimensional and R value stability over a large period of time (approximately 20 years minimum).
5. Tongue and groove configuration to reduce cold spots and water movement between the insulation sheets.
6. Low cost, presently available and easily handled.

For inside placement, the environmental conditions change; hence, the important characteristics become:

1. High resistance to fire or the production of poisonous fumes during fire.

2. Same as 4 above.

3. Same as 6 above.

4. High R-value per inch of thickness to minimize interior space loss.

One should note that low cost appears on both charts. Specifically:

low cost = maximum long term R-value per dollar spent

Hence, an insulation that has a high first cost, but constant long term properties, may be the most economical product. The table below lists four major insulations and several of their properties. This data was taken from a recent summary paper of the field performance of various insulations by Dechow and Epstein of the DOW Chemical Company (see bibliography). Because little compiled data exists on this subject, the numbers given are based on a very limited number of samples. The data is also based on unfaced insulation with no external water protection.

	initial[1] R-value per inch	final[2] R-value per inch	cost[3] per board ft.	R-value[4] per $
extruded polystyrene	5.0	4.54	.16	28.38*
bead board	4.0	2.85	.11	25.91
polyurethane	6.89	2.98[5]	.19	15.68
fiberglass	3.5 (for comparison)			

1. Initial value for insulation given in study.
2. Final R-value for 10 years exposure to the ground, from study.
3. Installed cost per board foot based on 1977 Dodge & Means Cost Handbooks.
4. Cost efficiency using the 10 year R-value (for illustration since prices vary).
5. This data for polyurethane has been questioned as being unusually low.

5-7 R-value per dollar

One should note the various initial/final R-value differences. Beadboard and polyurethane both have a greater loss of R-value than extruded polystyrene in long exposure to the ground because they are more permeable and absorb more water. Clearly, protection for beadboard and polyurethane against moisture pickup from the ground should be considered. A faced insulation or external waterproofing can be used to improve the long term thermal performance.

With respect to the requirements listed for outside insulation placement, extruded polystyrene is the most suitable since it satisfies all the criteria; hence, it is highly recommended. However, with respect to the properties required for inside insulation placement, fiberglass may be the most suitable because of its low cost and the reduced performance requirements for internal insulation. As stated previously placing insulation inside is not recommended for earth sheltered housing. Apart from the disadvantages mentioned above, insulation placed inside a waterproof or vapor proof layer will cause the surface temperature of that layer to cool substantially since it will be on the cold side of the insulation. Any water vapor permeating outwards through the insulation will then meet a cold impermeable layer which may cause condensation, an increase in the moisture content of the insulation and hence a drop in the effective long term R value.

6 public policy issues

- building codes
- h.u.d. minimum property standards
- o.s.h.a. standards
- legal aspects
- zoning ordinances
- financial aspects

introduction

Few houses anymore are built in remote areas where individual buildings will have little or no effect on the surrounding land. As houses and buildings have been built closer together and forming larger and larger cities, many problems have arisen from unrestricted building design. These concerns have included, for example, sanitation conditions, percolation of sewage into drinking water supplies, the drastic changing of surface water run-off conditions, the danger of fire spreading from one building to those around, and the property value impact of a tumbledown shack erected amongst more elaborate houses. Out of these problems and the need for more regulations have grown the present building codes, property law and zoning ordinances.

The building codes regulate the standards of design and construction for individual buildings including standards for mechanical, electrical and plumbing work as well as the building structure itself. These codes carry the force of law where they have been adopted by states or local communities. A closely allied set of building standards is the Department of Housing and Urban Development (H.U.D.) Minimum Property Standards. These do not carry the force of law but are the basis of acceptance for government underwritten loans on housing and are hence important to consider for present or future financing of a house, and its resale potential. The Occupational Safety and Health Administration (O.S.H.A.) also has regulations which can affect housing construction. They are concerned with the safety of the workmen during construction and do have the force of law.

Legal problems refer mostly to problems of liability to surrounding properties resulting from house construction. The legal aspects overlap into both the code and zoning areas since code or zoning laws may legislate some parts of the construction procedure to help avoid legal problems.

Zoning ordinances do not deal with the construction details but instead control the type, size and setting of buildings within an area to preserve a desired neighborhood character. They can vary greatly from one locality to another as opposed to code provisions which are very similar even in different parts of the country.

Financial aspects include the taxation, insurance and home financing aspects of constructing an earth sheltered home. Home financing, in particular, has always been a problem in the acceptance of innovative housing ideas.

building codes

Building codes are only applicable where adopted by local or state law. This usually applies to all urban areas and most suburban areas. As discussed above, building codes are particularly important for these areas because what one person builds on his property does affect not only his own safety and the safety of those who may later purchase the property, but also the safety of the people around him. The building code is mainly concerned with life safety, health, public welfare, and the protection of property.

Several nationally recognized model building codes exist, (see reference information). These model codes are prepared in a form ready for adoption by states or local communities. In Minnesota, for example, the Uniform Building Code (UBC) has been adopted with certain modifications to form the State Building Code and this State Building Code has recently been extended in its area of coverage to cover the entire State. Thus, rural areas which formerly did not come under any code restrictions are now automatically subject to the State Building Code. Since this report is primarily applicable to Minnesota, only the Uniform Building Code provisions (as modified by the State) will be considered. The Code Organizations listed above have combined to produce a document called the One and Two Family Dwelling Code which is compatible with the requirements of all the national model codes. The major provisions affecting earth sheltered housing, i.e. those affecting light, ventilation and exit provisions are almost identical to the UBC provisions. This indicates that the information given here will be more generally applicable than just in UBC areas. The International Conference of Building Officials (I.C.B.O.) which publishes the Uniform Building Codes also publishes companion codes and manuals. (Again see reference information).

Building codes generally have two goals:

1. To specify performance criteria that buildings must meet, i.e. buildings must be structurally safe, must provide adequate shelter for the occupants, must provide reasonable fire protection for the occupants, etc.
2. To give deemed-to-satisfy provisions which if met will automatically mean that the building is considered to meet the performance criteria of the building code.

The deemed-to-satisfy provisions, or prescriptive standards, are included to ease the burden on both the people who want to build conventional structures and on the code administrators. They actually reflect layouts and construction techniques that have proved suitable over a long period of time and hence are automatically recognized to meet the performance intent of the building code. In this way,a prospective builder can know when the plans for his building are produced that, if he follows the detailed provisions to the letter, he will have a structure which will be acceptable to the building code officials and also should be a sound and workable structure. The code administrator's job is eased because, if the plans meet the specific requirements, he is absolved of the responsibility of determining whether or not the performance criteria are met.

Unfortunately, the distinctions between the two types of criteria are not always clear. Two examples which have some bearing on earth sheltered housing and which might illustrate this can be given. In the first, a general criterion that any room receive an adequate amount of light and ventilation for its purpose is not given as a performance criterion in the code. There is instead only a prescriptive standard that exterior glazed openings with an area not less than one-tenth of floor area be provided in all habitable rooms (together with some minor modifications). This has great ramifications in the code administration because now the only requirement is the specific one. A local code official would be asking for trouble to allow a building directly contrary to code even though alternative provisions may seem reasonable to him. To do something other than is specified, would for this case, probably require a code variance which must be granted by a Board of Appeals and could become an involved and lengthy procedure.

A second example would be roof construction provisions. Section 320l of the Uniform Building Code states that roofs shall be as specified in this Code and as otherwise required by this Chapter. Obviously the code does not specify an earth sheltered roof as an alternate, so the acceptance of the roof must be based on Section l06 Alternate Materials and Methods of Construction. This general section states, "The provisions of this Code are not intended to prevent the use of any material or method of construction not specifically prescribed by this Code, provided such alternate has been approved." The onus is clearly on the builder to prove that his construction is at least the equivalent of that prescribed in this Code in quality, strength, effectiveness, fire resistance, durability and safety. In this case, approving different materials does not directly go against a code provision and the building official can use his own expertise and judgement. For the case of an earth sheltered roof, it would be normal for the building inspector to require evidence of a structural analysis of the load bearing capacity of the roof members by a competent engineer.

There has been a long standing argument between performance and prescriptive standard proponents. Prescriptive standards have advantages in allowing easier enforcement. Both parties know what is required of them and personalities and reputations enter into the situation very little. Prescriptive standards work best for conventional buildings where little design effort is used. In unconventional designs, however, prescriptive standards may be unnecessarily restrictive or even out of place. They allow very little ingenuity in meeting the intent of the code. Performance standards overcome these objections by specifying the performance of the end product, not how it is to be achieved. In performance standards, however, local code officials are required to make judgments on the final performance of systems that may be outside their realm of competence. The fact that they are allowed to make judgments removes some of the rigidity of the code but also introduces a greater element of uncertainty. An uncooperative local official would have more leeway for obstruction than under a prescriptive code. Involving registered architects or engineers should help in situations like these. Such professionals can provide expert data or opinions to the code official, and their involvement will help relieve the code official of some of his responsibility in the decision.

Detailed building code regulations which have particular relevance to earth sheltered housing are presented in Appendix D. These code requirements are outlined with little direct comment on the applicability of the provisions to earth sheltered housing. The major concerns affecting the layout of earth sheltered houses can be summarized as:

1. Habitable Rooms must have glazed areas greater than one-tenth of the floor area.

2. Opening windows of at least one-twentieth of the floor area are required unless mechanical ventilation is provided.

3. Sleeping rooms must have a window or door connecting directly to the outside.

The first requirement has been established for a long time. It is a specific requirement and one that individual Code officials will be reluctant to allow variances on. No changes in this provision were made when the recent energy code revisions were made, and apparently no moves are presently under consideration to change it, even though the H.U.D. office of Underwriting Standards is actively considering reducing the minimum window area to 8% of the floor area in their minimum property standards.

This provision can be met without undue difficulty by earth sheltered houses of many different designs, but at the same time it does represent a major design restraint. The differences in opinion on this provision between some underground housing enthusiasts and the code bodies do not generally occur in the living/dining/kitchen areas where almost everyone would agree natural light and a view outside are of great importance. The differences come in rooms like studies or dens, and bedrooms. Many people feel that such rooms have no need for natural light and,therefore, they can be put at the back of a house plan adjacent to an earth wall rather than against an exposed wall. Having to place such rooms against an exposed wall causes layout problems and probably will expose a greater surface area of the house to the elements. Of the three major requirements listed, this requirement perhaps has the least basis in necessity and there would seem to be no real reason why the requirements could not be more flexible than at present to allow an individual some measure of control in where he does or doesn't want natural light.

The second requirement for natural/mechanical ventilation appears quite reasonable and does allow an individual some flexibility in meeting this provision. Mechanical ventilation is not hard to provide. In fact, with forced air heating or cooling systems, the necessary circulating system will already be included in the design. Natural ventilation of earth sheltered houses is also quite possible and hence the designer can assess the trade-offs between the different approaches discussed earlier in the report.

The third requirement for sleeping room exits presents some of the most unclear questions. The intent behind the requirement is extremely important in any house design; namely, that if a fire should start in a part of the house other than the bedrooms, sleeping occupants once awakened should have a clear means of escape directly to the outside without having to go into a possibly smoke or fire filled part of the house. Because a bedroom may not always contain young healthy people, further restrictions are placed on the minimum size of the window and the maximum height from the ground. It should be noted that an exterior window is not always regarded as direct means of escape. In second story windows or above, it is a place of refuge with access to outside air and from which the occupant can be rescued. Clearly the design for smoke and fire protection to the occupants must be a strong consideration in the design of a house with limited exterior openings.

The main argument with the code requirements would be that there appear to be other ways of giving equivalent protection. This is especially true for families where the parents will almost always enter whatever other parts of the house are necessary to reach their children to lead them to safety.

Again there is really no problem in meeting the existing code requirements with an earth sheltered house. In fact, all the plans developed in these guidelines do meet the existing code requirements. These requirements are, however, like the minimum window area requirements, a strong controller of design.

Some alternatives to the code requirements have been proposed and accepted in other parts of the country:

I. Andy Davis' house in Armington, Illinois, is all concrete construction with an exposed rock finish. There are no furring strips, wallboard, paneling, etc. and hence the only combustibles are the interior furnishings. It should be noted here, however, that in terms of life safety furnishings alone are sufficient to give off large quantities of toxic fumes when burning. A plan of his basic house is shown alongside. He has apparently been able to obtain code approval in several states using one or other of the following techniques in lieu of the normal window opening requirement:

- A smoke detector outside each bedroom that has no direct access to the outside — used for his basic house.
- An opening skylight in the bedroom with a rung ladder cast into the exterior wall — used when combustible interior partitions are incorporated.

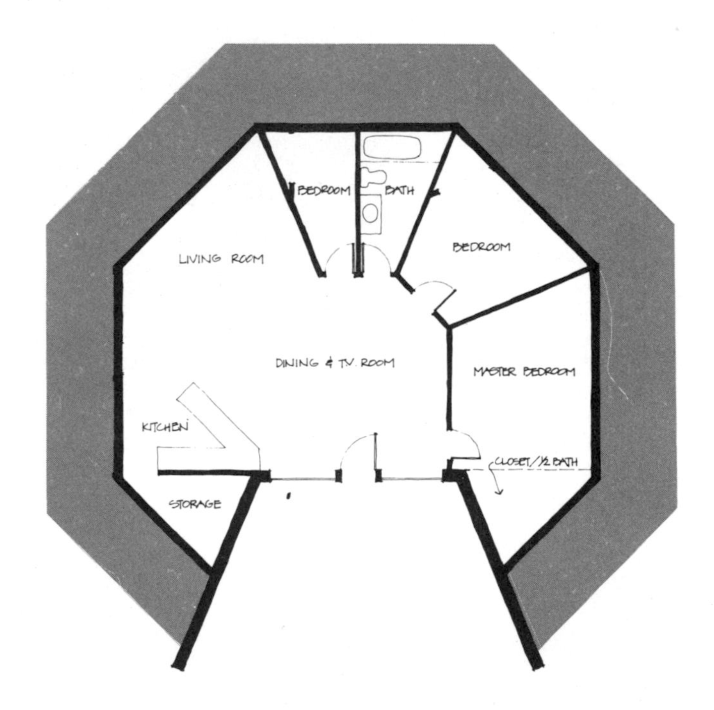

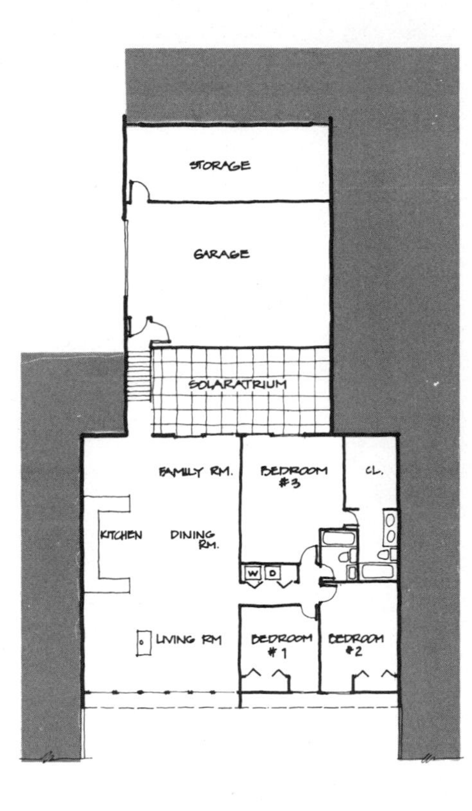

- A back corridor around the bedrooms leading to a back exit. The corridor should have fire rated doors opening onto it so that they will hold the passage of smoke and fire long enough for the corridor to be used as a means of exit — again used when combustible interior partitions are incorporated.

2. A house being sold by Solar-Earth Energy Homes of Ohio (see adjacent floor plan) has bedroom #3 which has no direct exterior access, but does have two alternate escape routes - one through the house and another across a covered atrium to an exterior door.

With little track record of widespread use, it is probably too early to say whether smoke detectors alone give the equivalent protection of having a direct exterior access. Since smoke detectors are now required in new homes anyway, it will be almost impossible to prove that smoke detectors alone provide more protection than the present code which requires both. Smoke detectors alone may well prove, however, to provide better protection for families than the old exterior opening requirements with no smoke detector.

An opening skylight with a rung ladder would be a very acceptable means of escape for most people. For very old, handicapped or even temporarily injured people, it would not be as feasible. These people would also, however, have extreme difficulty in climbing out of a 20″ wide by 24″ high window 44″ from the ground. This leads into the wider question of whether a person is free to design a home for his own needs, or whether he must design it with future ownership by others in mind. From a code administrator's point of view, the answer must be fairly clear — most homes do change hands at regular intervals and the particular needs of the first occupants should not be allowed to create an unsafe condition for later owners. From a potential home builders point of view, the answer must appear equally clear— why should he have to build his house to satisfy regulations that have no bearing on his particular needs? Fortunately, the code for the most part simply represents good design practice which has been found to perform satisfactorily over a long span of time. This question cannot be resolved here, and the potential builder must be aware that these are the sorts of objections that he must overcome to be granted a variance.

The back corridor around the bedrooms leading directly to a separate exit does appear to offer equivalent protection to the opening window requirement. In fact, for families or old people it could offer greater protection. For families, the other bedrooms can be accessed from the corridor and there would be no need to enter the main body of the house. For old people or handicapped persons, exits through doors would be far easier to traverse than a window exit. The provision of 2 alternate exits is more in line with the life safety requirements of larger buildings. Similarly, the two alternate escape routes of the Solar-Earth Energy Homes plan appear to provide a satisfactory life safety provision. If such a scheme is to be presented as an alternate to the code, the plan should be clear that the alternate escape route will be through an area which will be unobstructed, i.e. it should not pass through a garage, storage room, etc. The main disadvantage of the second exit is cost of this provision for no extra useful space. This is mollified somewhat if the exit can be combined with another function such as the atrium.

A further technique which can be used to make the alternate escape route more acceptable is to arrange the mechanical ventilation system for the house (if provided) such that a positive ventilation pressure will exist in the escape corridor or atrium. This is the same technique used in larger buildings to provide smoke free escape routes during a fire. Because the air leaks from the corridor or atrium area into the rest of the house, no smoke will enter this area until the fire (by heating the air) creates a higher pressure in the rest of the house.

A note in connection with covered atriums is that windows from adjacent rooms opening onto the atrium cannot be automatically assumed to meet the natural light requirements of the code because such windows are not technically exterior windows. Some precedent for the acceptance of alternates in this area has been set by the approval of windows opening onto roofed porches.

Two other options can be considered in planning for exit and natural light requirements:

I. A small high window in a back bedroom may be desired which would allow maximum earth protection to this area of the house while still providing light, ventilation and an emergency exit. The maximum sill height of 44″ for an escape window does not allow placement of the window as high as might be otherwise considered.

   A permanent chest/seat combination or raised floor built adjacent to this wall under the window can probably be used to allow the window sill to be raised to within 44″ of the top of the seat or raised floor. The important considerations here will be that the construction should be permanent and sturdy. It may be required that minimum standard steps be provided to walk up onto this raised level. This will depend on how the rescue or escape elements are weighed. For instance, it is hard to see how anyone who is expected to escape through a window 24″ high by 2O″ wide, 44″ off the ground would be unable without steps to climb onto a seat 24″ off the ground. When rescue is considered, however, there could be a slight advantage in the case of an old person who could be pulled to safety when standing adjacent to the window, but may not be able to be reached if he or she could not climb onto the seat. Such arguments can become very limited in scope, and similar "what-if" arguments could be cited to show that almost any code provision doesn't provide safety under certain conditions.

2. Rooms can be combined for natural light and ventilation requirements if the provisions in Sections l4O5 (a) are met. These provisions can be used effectively in open plan arrangements for the living room, dining room and kitchen. In rooms where some privacy may be required at certain times, but generally the room would be a part of adjacent rooms, moveable curtaining across the required opening would appear to be a reasonable compromise meeting most of the intent of the code. Rooms that could be considered for this sort of treatment would be a den, a bedroom in a cabin, or perhaps a guest sleeping area in a house. Code officials in Minnesota have indicated that they do not feel curtaining meets the intent of the Code as it stands. A major concern would be what type of curtaining was involved. Moveable but heavy partitions would have different implications from easily drawn curtains.

An attempt has been made to review the pertinent conditions of the Uniform Building Code with respect to earth sheltered housing design. This is intended to highlight the code problems faced when designing such houses. It does not cover, by any means, all of the code requirements pertaining to housing and it does not give the full wordings of the code. The respective codes should be consulted directly for more detailed guidance. Plumbing, Mechanical and Electrical Code requirements will not be reviewed as they have few implications for earth sheltered houses in general. Some problems can perhaps be anticipated in getting code approval for novel heating, cooling and thermal storage methods. As in all code questions, good documentation of the system and the assistance of competent professionals will be of help in trying to obtain a permit.

h.u.d. minimum property standards

The H.U.D. Minimum Property Standards is a guidance document intended to define the minimum level of acceptability of design and construction standards for the numerous programs of the Department of Housing and Urban Development. The standards are also used by the Federal Housing Administration for determining acceptability of houses for F.H.A. loans. If resale or ease of financing are major considerations in the planning of a house, then building in compliance with the H.U.D. standards as well as the building codes can be well worthwhile.

As when dealing with the building codes, only the provisions which appear to have particular relevance to earth sheltered housing will be discussed. Three of the stated aims of the standards are very much in sympathy with the earth sheltered construction movement. These are:

1. To provide improved design and construction standards based more on performance than has been true in the past, with appropriate flexibility to meet local conditions.
2. To encourage design innovations and improved building technologies, giving promise of increased quality and reduced costs.
3. To aid national and local efforts being made to improve the environmental factors of urban areas.

The particular standards discussed here will be for one and two family dwellings (1973 Edition). The standards are for sale by:

The Superintendent of Documents
U.S. Government Printing Office
Washington, DC 20402

The standards provide a large amount of data representing good practice in residential design and would serve as a valuable reference book.

A brief summary of the most pertinent standards is included in Appendix E. It should be emphasized that the H.U.D. standards are approximately 300 pages long and that only very brief highlights are given in the appendix. It should also be noted that meeting the H.U.D. minimum standards does not ensure that a particular house will be acceptable for financing. They will also be concerned with the general acceptability of the project to the public and this is where their concerns will tend to parallel those of private banks (discussed under financing).

In general the H.U.D. standards and the building codes are similar in their common requirements. Three important differences are noted below:

1. Mimimum glazing requirements may soon be reduced to 8% in the H.U.D. standards as opposed to l0% in the building code.

2. There are no optional mechanical ventilation provisions for living and dining rooms, bedrooms etc., in the H.U.D. standards.

3. Two separate means of escape from a bedroom are permitted in lieu of an exterior door or window.

The standards appear to have more leeway and encouragement for novel building systems and planning than do the building codes. There should be few problems in meeting the H.U.D. standards with an earth sheltered building if the standards are considered in the building planning. It should be noted, however, that local offices can have some surprises; for instance, the Minneapolis office has a rule that basements must be more than 4 feet above the water table. The national standards, on the other hand, do permit space below the water table if walls and slabs are fully waterproofed.

o.s.h.a. standards

These standards are primarily concerned with the safety of people at work and hence have little relevance to the design of houses. They do, however, have relevance and carry the force of law in house construction and therefore, will be of concern to anyone who employs other people in the construction of a house. There is a special section on construction safety standards, and the standards can be obtained by contacting the local Occupational Safety and Health Administration which is a division of the U.S. Department of Labor.

The most relevant sections for earth sheltered housing are on the temporary protection of banks and trenches during construction. Large numbers of people are killed each year with tragic regularity by the cave-ins of banks and trenches. These can be quite sudden and in deep trenches are usually fatal. The O.S.H.A. standards require that banks and trenches greater than 5 feet in depth shall be sloped or shored. They further require that trenches of lesser depth be shored when hazardous conditions exist. Moveable trench boxes or shields may be used in lieu of shoring. Ladders are required for trenches greater than 4 feet in depth with a maximum travel distance to a ladder of 25 feet. The standards also give detailed shoring provisions and angles of repose for various types of ground. Major changes are expected in O.S.H.A. regulations shortly, however.

Few legal aspects of earth sheltered homes are any different from the legal liabilities of constructing and owning any home. In large scale underground buildings where different uses of the same piece of land by different concerns are a possibility, the legal problems can be quite complex. For single family dwellings on their own individual lot, however, such problems do not arise except with utility easements or mining rights which are common to all types of houses.

Five legal questions with particular significance for earth sheltered houses have been chosen for brief discussion. Some legal opinions were sought from the Minnesota Attorney General's Office, but the Attorney General's Office can only respond to opinions requested by local government officials who need assistance or by a legislative body. Also, most legal questions are extremely dependent on the specific circumstances and general opinions can be of little value.

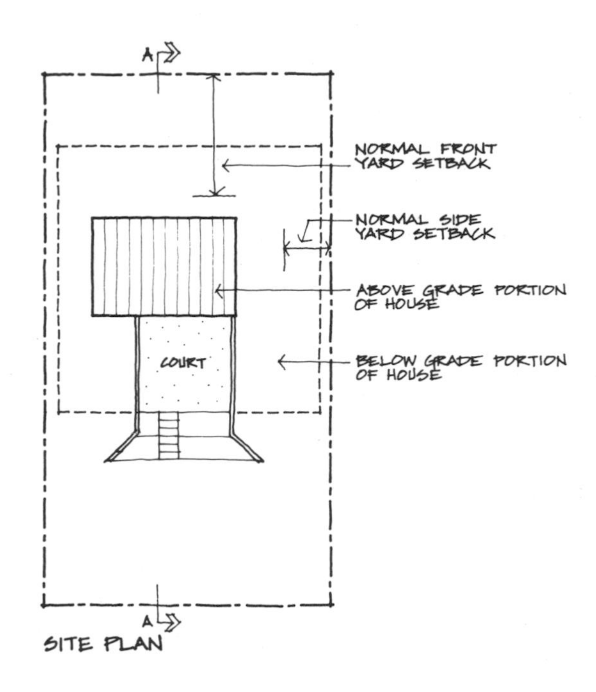

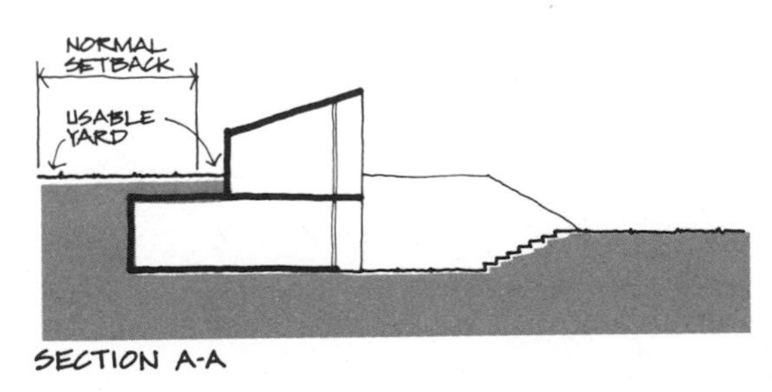

6-2 setbacks

setbacks

There has been some question as to whether set backs for completely underground portions of houses need follow the same requirements as for conventional houses. The major reasons for set backs are:

1. To provide fire department access,(mainly applies to side set backs to permit fire department access to the rear of a building).
2. To ensure the proper light and ventilation to required windows facing a property line.
3. For neighborhood aesthetics.
4. To allow maintenance of the wall exterior from the owner's own property.

On a tight site there may be the desire on the part of the owner to use a greater portion of his site than would normally be permitted. A completely underground portion of a house extending into normal set backs would appear to interfere very little with the first three aims listed above. The maintenance access problem has already been addressed in connection with the H.U.D. standards which do allow a windowless wall at the property line if permanent access for maintenance has been legally secured. The building code requires fire resistive construction within 3 feet of the property line but a concrete wall underground should certainly meet this requirement. The major objections to such a scheme will probably come then from the administrators of local zoning ordinances which also specify set backs. A general precedent will probably be determined by the first few test cases should this problem ever arise.

planned unit development

When several dwelling units are being built by one developer, it is possible for him to apply for status for the project as a planned unit development. This status allows the project as a whole to be considered in meeting open space requirements, etc. It also allows houses to be built with common party walls in the townhouse or condominium style, i.e. with zero side wall setback. Since many of the advantages of earth sheltered housing become more apparent when a number of units are considered together, the status of a planned unit development will make such projects doubly attractive. There will be a flexibility in layout and use of space that usually could not be achieved under conventional restrictions. Planned unit developments are administered locally and so may not be recognized in all areas. A check should be made with the local planning office as to the availability of such status and the minimum number of units to qualify.

effect of excavation on adjacent buildings

The majority of excavations for earth sheltered houses will not extend much deeper than the basement of conventional houses, and as such, the legal liability will not be greatly increased by expanding this type of construction. There will be occasions, however, when excavations must be made deeper than normal or extremely close to the property line. In these cases great care must be taken to protect the adjacent property from subsidence of ground into the excavation, or worse still, the actual subsidence of a nearby building.

The building codes already provide some protection against subsidence and some legal guidelines in this area. For instance, they regulate the distance of cuts and fills from the property line (U.B.C. Chapter 70). They also lay out the procedure for notification of the owner of the adjacent property and the respective responsibilities of the two owners for underpinning the foundations of an existing building. Though the building code states that the owner of the adjacent property is responsible for protective measures to his foundations to a depth of l2 feet, it cannot be assumed that the owner causing the excavation will not be responsible for any damage that occurs because of his excavation, even if the adjacent owner has not taken any protective measures. The only mitigations here might be that if the adjacent owner is properly notified and refuses to allow the underpinning of his building, he may not be able to recover the full costs of any damage caused. This would also apply if the adjacent owner's building was an accident waiting to happen. An actual example of this latter condition was a house built on poorly compacted fill. Construction traffic on the street in front of the house triggered a collapse of the ground adjacent to the house into voids existing in the fill. The construction traffic vibrations no doubt triggered the movement but the subsidence would probably have occurred sooner or later anyway because the fill was in an unstable condition. On projects requiring large areas of deep excavations such as subway construction, a great deal of effort is put into deciding which adjacent buildings must be underpinned and surveying all the adjacent buildings for existing cracks and structural defects. If such a survey is not made, the owner is open to fraudulent claims of damage against which he will have to prove that he is not liable.

disturbance of water flow patterns

This is similar to the above question of disturbing the ground conditions of adjacent property. Again, the same liabilities apply to any house construction but because earth sheltered housing often requires more landscaping, this question is highlighted here. Legal liabilities could occur both by diverting water away from an adjacent property that used to benefit from the water and by diverting water onto an adjacent property causing drainage problems. In the latter case, concentrated run-offs in particular should be avoided since these can cause flooding and erosion during heavy rains.

solar rights

The solar rights issue is basically whether people building solar collectors or passive solar homes should be protected against the future shading of part of their solar collection area by building construction or vegetation growth on adjacent properties. If it is agreed that solar rights protection is required, the next question is how to write legislation that will protect reasonable solar rights without infringing unreasonably on the use to which an adjacent owner can put his land and without being too complicated for widespread administration.

For an earth sheltered house, the solar collection area, whether active collectors or windows for passive collection, is likely to be lower in elevation with respect to the surrounding ground level than for conventional housing. This magnifies the solar rights problems on small sites because limits on adjacent buildings and vegetation must be more severe to give the same degree of protection against shading.

At least two States have passed some solar rights legislation into law. Colorado has passed a measure which establishes a procedure for setting up permanent solar easements in the same way that easements are negotiated for utilities. Such easements could be written without the State law but the law will standardize the easements and probably make them easier to obtain. New Mexico has passed legislation based on western water law which essentially provides that the benefits that an existing owner enjoys should not be interfered with by new construction on adjacent land. The Minnesota Legislature considered solar rights measures in its last session but no legislation was completed. The Minnesota Energy Agency is currently drafting new legislation in this area which will be considered in forthcoming sessions.

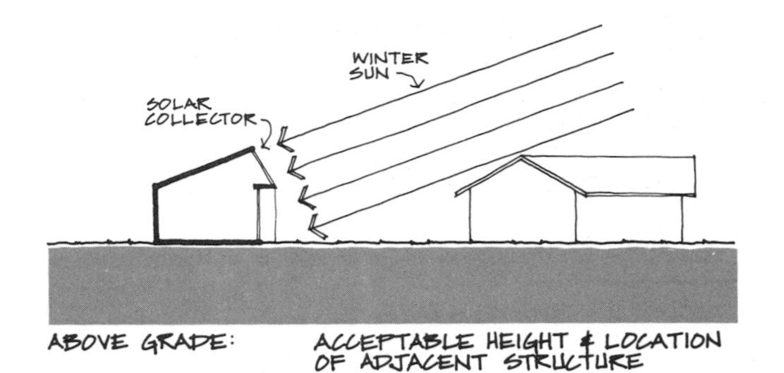

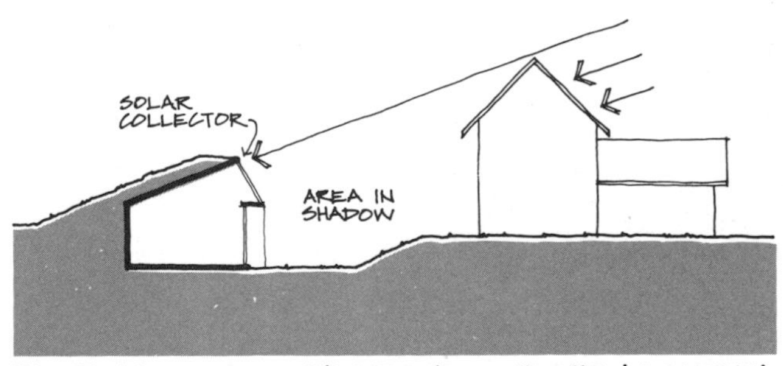

6-3 shading problems

Zoning ordinances and similar planning restrictions are locally adopted and administered rules that control the type, size and setting of buildings within a particular area. They are designed primarily to separate incompatible uses of land as much as possible — trying to avoid for instance, a noisy factory being constructed in a quiet residential neighborhood. Zoning ordinances are also used to control the character of single use neighborhoods. The latter type of ordinance has the most implications for earth sheltered housing and will be the only type looked at here.

Because zoning ordinances are local in origin, they can vary greatly from place to place, and the following discussions are only a sample of what restrictions can exist. A number of outlying communities around the Twin Cities area were contacted to see whether zoning ordinances might prohibit or severely restrict earth sheltered housing. The general reaction of the zoning administrators was that if the house would meet the building code, there should be no special problems with the zoning ordinance.

So called "Basement Ordinances" had been thought to be a possible problem at the beginning of the study. These were supposed to have grown out of a fairly common practice after the war of getting financing to build a basement and adding the rest of the home at a later date. Many people did not complete the remainder of their houses for substantial periods of time and such unfinished houses were considered to be an unsightly blight in the neighborhood. Only one of the communities contacted, however, even had such an ordinance. This was the City of Maple Grove which has an ordinance prohibiting living in a basement (as defined in the Uniform Building Code). The building inspector for Maple Grove had already anticipated this question with respect to earth sheltered houses and said that this ordinance would be reviewed if an earth sheltered house was presented to them which would not pass under the existing definitions.

Most zoning ordinances have some restrictions on the minimum lot size or minimum square footage, and may require dwelling units to have a garage. The City of Elk River, for instance, in its square footage requirements has a minimum square footage of 750 square feet for a single level house. For a two level house, the main level must be 720 square feet with no requirement for the lower level.

For a two level earth sheltered house, the building inspector interpreted this to mean that 720 square feet would be required on whichever was the main level with no requirement for the other level.

The City of Forest Lake zoning administrator foresaw no special problems in granting permits for earth sheltered houses except in some of the very low lying areas where flooding and high water tables would be a problem.

Set backs may be controlled under zoning ordinances, and as discussed under legal aspects, there is a good case for reduced set backs for the underground portions of a house. Also discussed under legal aspects, were planned unit developments which can be used in certain areas to negotiate the planning restrictions for a whole development rather than having to meet the individual criteria house by house.

In summary, a very positive attitude was found amongst the city officials contacted. There appeared to be very few zoning restrictions which will unfairly hamper earth sheltered housing. Where such a restriction did exist, the city official was ready to consider changing the ordinance. Individuals applying for permits in their own particular area may not always find everyone so helpful. Clear drawings of the proposed house, together with a proper engineering analysis and the proposed details of construction, will usually help more than a thumbnail sketch on the back of an envelope. This shouldn't be construed to mean that a preliminary meeting with the local building inspector is not a good idea. Such a meeting can be very helpful. It gives the inspector a chance to voice his concerns about the house concept before he is put in the position of having to grant or deny a permit. It should be made clear at such a meeting that formal plans and application will be presented later and that the official is not being asked to prejudge the applications before he has all of the details. The meeting should be merely to acquaint the official with the project, get his overall impressions of the idea and to give him some input into the planning of the project.

taxation

The method of assessment for tax purposes of new construction in Minnesota was investigated to find out the manner in which earth sheltered homes would be affected by assessment rules. The Minnesota Department of Revenue was contacted and they gave the following information on tax assessment:

1. The assessment is based on the square footage of the house with adjustments for quality of construction, roof type, etc.
2. The cost factors used typically allow for basement space (this is not included in the square footage used above). If there is no basement as normally would be the case for an earth sheltered house, an allowance for space normally associated with a basement would be made, i.e. storage space, furnace room, etc. would not be included in the square footage of the house for assessment purposes.
3. In Minnesota, the assessed value must be within l0 percent of the average level of assessment in the area. This is to prevent inequities between the taxes of new construction and existing housing which may be undervalued for tax purposes. For instance, if existing housing with a market value of $60,000 is currently assessed at $50,000 , then a new house worth $90,000 would be assessed at close to five-sixths of $90,000 , i.e. $75,000.

From the discussions with the Department of Revenue, it appears that earth sheltered houses should not suffer from any peculiarities of tax assessment.

insurance

It was felt at the outset of the study that earth sheltered homes could have some advantages over conventional homes for insurance purposes. Upon closer study, however, it does not appear that this will be a reality, at least in the near future.

Insurance of buildings is based on ratings given to the particular type of construction, location, occupancy, etc. for the building under consideration. These ratings can be made by the insurance companies themselves but in over 50 percent of the States, the rate making body is the Insurance Services Office. This office prepares ratings and furnishes data for use by the individual insurance companies but an individual company is still free to adjust the ratings or prepare their own if they so desire.

The reason that there is little possibility of a rate reduction for earth sheltered homes is that all housing insurance is class rated. This means essentially that it is not worth the insurance company's time to inspect each home individually for fire and other hazards. The losses from any particular house are minute in relation to the total pool of houses insured and the insurance companies do not feel there is sufficient difference in risk between different types of houses to warrant individual treatment. There are, however, some modifiers which affect the homeowner's premium and these will be outlined below.

The class of insurance coverage desired is the first determinant of the insurance rate. There are five major classes of insurance for residential buildings:

1. Fire and extended coverage — Protects from fire, windstorm and hail loss, etc.
2. Theft — Protects from loss by theft of goods from the property.
3. Public Liability — Protects from liability to persons injured on the property.
4. Tenants Form — Protects only contents.
5. Broad Form — General policy including coverage under l, 2 and 3, and providing some additional coverages.

The insurance rate for the type of coverage is then modified by the following factors which can be grouped into the three major determinants of rate which apply to any building:

1. Type of Construction — As mentioned above, houses are not graded by construction in detail. There may be a simple split between frame and masonry construction in some ratings.
2. Type of Occupancy — The only major determinants here will be information as to the number of families living in the dwelling and whether the building is owner or non-owner occupied.
3. Exterior Grading — This refers to the quality of public fire protection; for

instance, water availability, water pressure, distance to fire station etc. There are ten grades of protection, with Grade l0 having no protection and Grade I the best protection. The Cities of Minneapolis and St. Paul, for example, are considered Grade 3.

Insurance rates are also adjusted yearly by a loss ratio which indicates whether, for an identified class of risk, the rates have proved to be too low or too high in the past. Because earth sheltered houses will be a tiny group within the vast pool of housing for some time to come, any better insurance risk for earth sheltered houses will not become apparent unless the insurance companies decide to review their records. It is likely to be many years before:

1. Earth sheltered housing forms a large enough group to warrant separate consideration.
2. A sufficiently long and broad data base is available to allow a meaningful statistical analysis.
3. The insurance companies see any benefit to differentiating between earth sheltered and conventional houses.

Would the rate be likely to go down if earth sheltered houses were made a class of their own? To answer this question, a brief look can be taken at the present insurance system for commerical buildings which are rated on more of an individual basis. Again the major determinants are type of construction, occupancy and exterior grading, but for commercial buildings, the details of the construction and occupancy are looked at far more closely.

For fire rating, the basic rate is determined by the resistance to fire of the basic building structure including frame and walls. The construction class is rated for insurance purposes from Class I (wood frame) to Class 6 (2 hour fire rating or better).

Within the type of construction modifier, there are also secondary modifications for area and height and internal finishes. In general, the adjustment factor goes up both for space above the ground and below the ground. The major reason for the factor going up for the fire insurance of below grade space is the difficulty of fighting fires underground. Although the nature of the building restricts air flow to a fire, this can be more of a hindrance than a help. There will be usually sufficient oxygen within the building to allow the fire to build up at a normal rate. By the time the fire has reached such proportions that it needs more outside air, it will already be a serious fire and have filled the building with smoke and fumes. The restriction of air to the fire will actually cause more lethal concentrations of carbon monoxide and other toxic gases to occur. The fire fighters have only a few points of access to the fire area and cannot use their normal venting techniques.

There are several points to note about this apparently gloomy picture. The first is that a basement for insurance purposes is defined as being more than 50 percent below grade, unless it has horizontal access at grade level on one or more sides. Also there are sprinkler systems and automatic smoke vents which can ameliorate these drawbacks to large scale underground space. It must further be remembered that although the adjustment factor goes up for underground space, this is only a modifier on the construction class and since underground construction will generally be in concrete and steel the insurance rate will still be much lower than for a combustible structure above ground.

Along with the basic construction rate, the other factors assessed are type of occupancy, the exposure risk from fires in neighboring buildings, the degree of internal fire protection such as sprinklers and the quality of public fire protection. For buildings which have sprinkler systems (and most large underground buildings are required to have sprinkler systems), the public fire protection grading does not apply.

For a normal insurance company using the I.S.O. rating then, the fire insurance rate for an underground building will be slightly higher than for a building of equivalent construction above ground. If the building has at grade access, then the rate should be no different. The rate will, however, tend to be very low because the construction will be very fire resistant and sprinklers which can dramatically reduce fire insurance rates are usually required.

The other facets of insurance are extended coverage, vandalism protection, liability etc. Liability coverage would not be expected to change significantly from an earth sheltered building to a conventional building. Vandalism and particularly extended coverage which includes windstorm and hail damage etc. could be expected to be considerably cheaper for earth sheltered buildings. Unfortunately, however, it appears that extended coverage and vandalism only form a very small percentage of a normal premium.

Thus, even though the individual coverages may be substantially cheaper, the effect on the overall premium will be small. Much greater reductions would be apparent from a change in construction type and/or from adding sprinklers. For instance, a warehouse of incombustible construction could have insurance rates drop by 75 percent simply by the addition of sprinklers. Even higher rate reductions would result by changing from an unsprinkled building of combustible construction to a sprinkled non-combustible building.

Summarizing the implications for the insurance of earth sheltered housing, it would appear that insurance rates are unlikely to differ at all in the near future but eventually could drop a substantial amount. This would be mainly because the type of construction is essentially fireproof (wood basements excluded) and not particularly because the structure is underground. Extended coverage and vandalism insurance should also drop substantially but these only represent a small percentage of a broad form premium. Many of the concerns about fire in large underground buildings will simply not apply to most earth sheltered houses. There will be access at grade to at least one side in most houses, there will be no parts of the house that are remote from an escape route, and if a fire does occur, the basic structure will probably remain intact.

home financing

The importance of the availability of normal financing for earth sheltered homes cannot be overstated. If such financing is not widely available, then earth sheltered housing will never reach more than a tiny percent of the housing market. To illustrate the problems of obtaining financing for unconventional houses, a paper by Dean Manson, first presented at the Conference on Alternatives in Energy Conservation: The Use of Earth Covered Buildings in July 1975, is reprinted in Appendix F. The reprinted article is a slightly revised version from the journal "Underground Space" Volume I Number 2. Also, please note that Dean Manson is no longer with Southern Methodist University.

Dean Manson's article illustrates the problems of obtaining financing for earth sheltered houses. The reactions of individual lenders, however, have not been all bad. It appears, in fact, that reactions can vary greatly from one lender to another. In the questionnaire that accompanied the request for permission to publish the details of the houses in Part B of this report, the owners or architects all responded that they had not had difficulty in financing the houses. Many were no doubt such strong borrowers that, as Dean Manson pointed out, the actual structure in question really didn't matter. For others, the down payments may have been high enough to give security to the lender. Nevertheless, an individual like Andy Davis was able to obtain financing from a local bank without difficulty and without any special strings. As he said in his interview for Mother Earth News, "... our banker was either smarter than the average banker or more willing to take a risk on a new idea."

Correspondence from other potential earth sheltered home builders has not been so encouraging. A letter from Duluth indicated that the local office of the FHA and a local lending institution (reputed to be the most liberal in the area) would have nothing to do with the financing once an underground house was mentioned. The loan officer continually referred to the idea as a novelty and fad and said it would not have good resale potential. Other letters have expressed some similar reactions.

The situation should steadily improve with time since more and more people are building earth sheltered homes and are obtaining financing. When these homes start to be resold, then the question of resale potential will become a matter of record rather than conjecture. The reactions are also different at different levels in the organizations that have responsibility for home mortgages. The writer from Duluth indicated that the local office of the FHA did not see any possibility of FHA financing. In a recent meeting in Washington, however, with some high level officials of HUD, National Science Foundation and the National Bureau of Standards, the response to the idea of earth sheltered housing was very well received. The difference is in perspective of the people involved. The government agencies are being pressed to increase the energy conservation measures in housing and to reduce the ever increasing problems of urban sprawl. They are also being pressured not to make new regulations that will cause the cost of single family homes to rise any faster than they are now. The idea of earth sheltered housing then should be, and appears in fact to be, quite appealing to them. To the individual loan officer at the other end of the chain of command, the picture is quite different. He is responsible for making safe loans within the guidelines of the institution to which he belongs. He is not charged with general social responsibilities. An unconventional loan application must represent unwelcome problems to him. Without any specific programs or directives to justify his position, he will be in trouble if he approves the loan and it turns out badly. He also will get no thanks for rejecting the loan if the potential borrower successfully appeals to his superiors.

In addition to the meeting with HUD officials in Washington, local contacts and presentations have been made with key personnel in the State FHA office, F & M Savings Bank, First National Bank of Minneapolis and the Northwestern Bank and Banco Corporation. The response was again positive in all of these meetings.

If the HUD headquarters in Washington would establish a program to specifically underwrite earth sheltered homes, or even if they were just to issue a directive that earth sheltered houses should not be considered a novelty, and that if they met the minimum property standards they should be considered for standard HUD or FHA loans, a rapid change in the financing picture would occur. Along these lines, the national HUD office has already indicated a willingness to consider earth sheltered homes under the Section 233 program, while it is conducting additional studies described below. Legislative interest will help a great deal here. In Minnesota such legislative interest has already helped immensely. The Legislative Commission on Minnesota Resources funded this study through the Minnesota Energy Agency. They have also established The Underground Space Center to function as a source of information and a research center for the many facets of underground space use. Senator Frank Knoll, was responsible for setting up the meetings with the HUD office in Washington as well as with the leaders of several local banks. He was also the author of the amendment to a housing bill which appropriated $500,000 to build and monitor some earth sheltered demonstration homes. This study is being administered by the Minnesota Housing Finance Agency and will intimately involve the agency with earth sheltered housing. Congressman Bruce Vento, from Minnesota, introduced an amendment to the National Energy Bill which was passed by the House of Representatives, August 5, 1977. The amendment requires that HUD must complete two detailed studies within the next three years regarding building code requirements, zoning and financing. Under the bill, a preliminary study will be prepared next year with a final report to be completed in late 1980. Congressman Vento's bill will necessitate that HUD become deeply involved with earth sheltered housing over the next three years and that recommendations for improvement in financing etc. will come from within the organization rather than outside. The message is clear — the legislature can really help in this area and they are interested in responding to the concerns of their constituents. Where there is an interest in earth sheltered housing, the local legislators should be kept informed, particularly if there are problems with code and financing.

There is little else to say at present about the financing of earth sheltered housing because the prospects appear to be in such a state of flux. As when dealing with code officials, a competent and confident presentation will help. Underground has a very negative connotation to many people so it is best to avoid that term. Walk out basements are well accepted as a desirable feature of a home even though such construction is partially earth sheltered.

The specific programs where help may be obtained with financing are listed and discussed briefly below:

national

FHA financing, HUD Section 203B loans

HUD minimum property requirements must be met in order to qualify. Obtain further information from the appropriate office.

HUD Section 233 loans

Under this program HUD will insure loans involving new technology which the commissioner deems to be significant in reducing housing costs or improving housing standards, quality, liveability, or durability or improving neighboring design. The chief advantage of the program is that the prescribed tests of economic soundness or acceptable risk are not applicable. There are other eligibility requirements, however, that must be met.

HUD Solar Heating and Cooling Demonstration Program

These grants under this program are made to builders, developers etc. rather than individuals for the additional cost incurred in installing solar equipment for heating and cooling.

DOE (formerly ERDA)

DOE has had plans for funding of earth sheltered building projects. These plans are still in formulation and little money can be expected to be available before the end of l978.

local

Minnesota Earth Sheltered Housing Demonstration Project

At the time of this writing the exact format of the project is not known, but it is expected that part of the money will be used for grants to builders and developers and part for completely state controlled projects.

Minnesota Housing Finance Agency

The Minnesota Housing Finance Agency is administering the above project and can be contacted for further details. They also make low interest loans to qualified individuals but are limited to make these loans on low cost housing.

7 illustrative designs

- design a
- design b
- energy use projections
- cost estimates

introduction

In order to illustrate many of the design considerations which are discussed throughout this study, two earth sheltered designs are presented in this section. The layouts presented and the type of construction used are intended to represent economically feasible, simply constructed houses. They are comparable to well built, well insulated above grade housing, although they should not be considered as overly luxurious. These designs are not intended to represent optimal configurations or prototypes, but simply to be used as a means of illustration. Three major aspects of these designs are discussed. First, the architectural design and site planning considerations are presented for the two designs. Then energy use projections based on methods used in section 3 of the report are shown and discussed. Finally, preliminary construction cost estimates (which were obtained from a local contractor) and various cost considerations are presented. The issue of initial costs and life cycle costs of a house are of primary importance to the development of earth sheltered housing.

design A

This one level design is a simple elevational plan since all of the windows are located on one exposed wall with earth placed against the remaining three walls. An 18 in. layer of earth is placed on the roof. Since the design was prepared for a hypothetical site, it is assumed that the window wall faces to the south which is the optimal orientation for passive solar collection. Space for a flat plate solar collector is also shown on the south facing elevation of the house. The garage which is linked to the house by an entry is located on the north side so that the public entry area and driveway can be separated from the private outdoor spaces. It should be noted that the house can be placed on a flat site as well as sloping site as shown in the drawings.

Design A represents a minimum program size for most families considering that the 1800 sq ft area includes mechanical and storage spaces normally located in the basement. In this one level elevational design, all of the living and sleeping spaces are placed along the south wall. The only spaces that do not require windows are the bath, mechanical, laundry and storage rooms which are placed along the north wall of the house. This arrangement creates a long narrow plan. Although the length of internal circulation in this house is acceptable, it would be excessive if a larger house were laid out in this manner. The sloping roof over the living, dining and kitchen spaces allows greater light penetration into these areas in the winter while the overhang and planter serve to diminish the unwanted solar radiation in the summer. Ventilation is also enhanced by the sloping roof since the warmer air can escape through the upper windows.

For purposes of energy and cost comparisons a variation of this design, referred to as Design A-2 was developed. In Design A-2 the earth cover remains placed against the exterior walls but is removed from the roof. In either case the garage roof may be earth covered or conventional.

section x-x

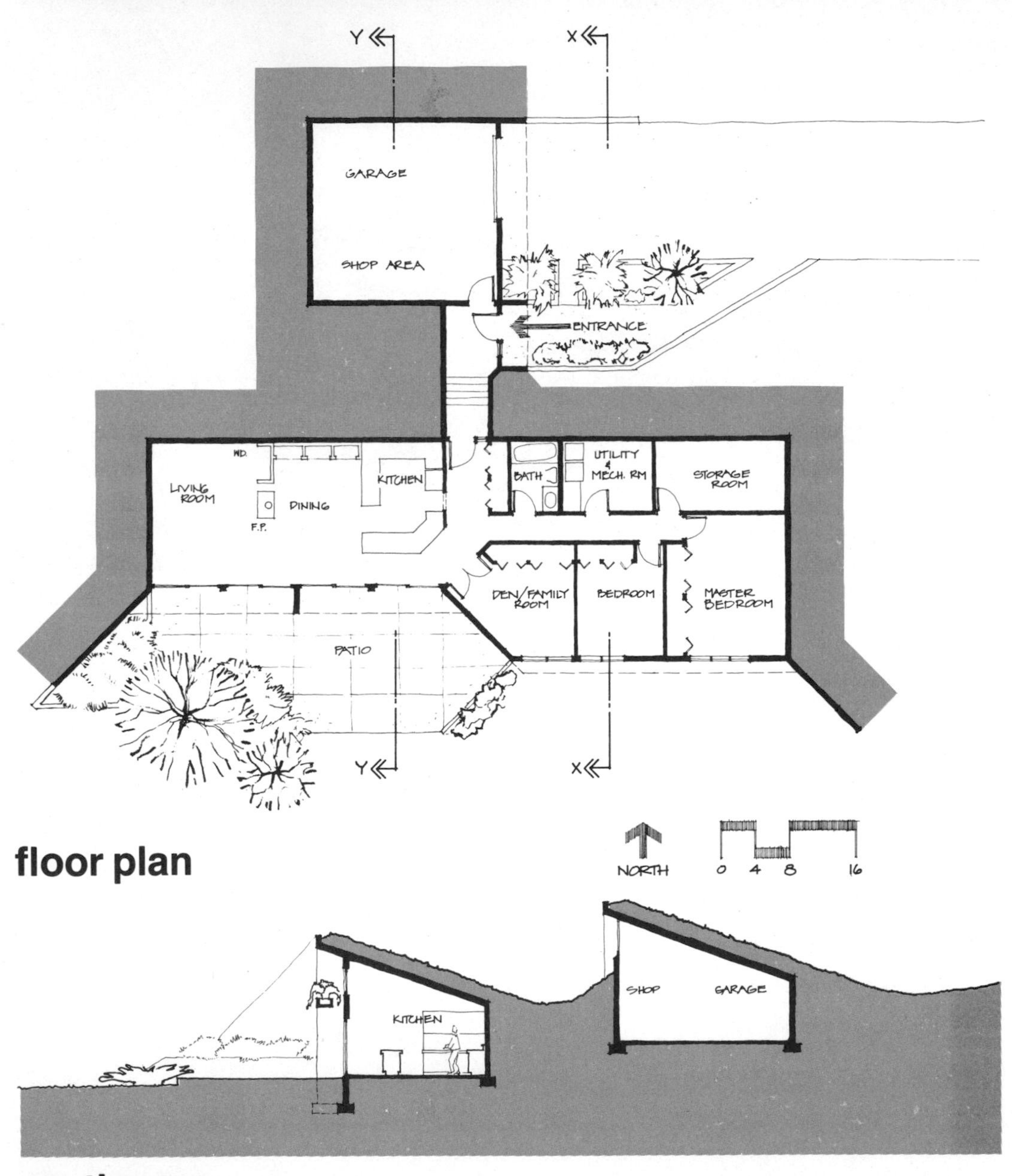

floor plan

section y-y

design by Carmody and Ellison

Similar to Design A, this large two level design is a simple elevational plan with all of the windows located on one exposed exterior wall. For maximum passive solar collection, it is assumed that the exposed elevation is oriented to the south. The remaining exterior walls are covered with earth over their entire two story height and 18 in. of earth is placed on the roof. In addition, the south wall of the lower level is partially earth covered. In this design, the entry occurs in the south elevation and the garage is located adjacent to the house with a separate entrance into the lower level. This arrangement does not provide as clear a separation of public entry and private outdoor spaces as an entry on the north side, as shown in Design A. However, this is overcome by separating the entry from the other outdoor spaces with level changes and landscape elements. A completely earth sheltered two level design such as this, is more suitable on a sloping site although it could be built on a flat site with substantial amount of fill.

The 2700 sq ft area of Design B is considerably larger than that of Design A. This larger program, which includes three bedrooms, a large family room and a workshop area, represents space requirements suitable for a larger family. In this layout the major living spaces are placed on the upper level which allows for a clear view of the outdoors and greater exposure to the sun. The bedrooms, storage and shop spaces are located on the lower level. On both levels the bathrooms, laundry, mechanical and storage spaces are located along the north wall away from the windows. A two level design is more compact than a one level scheme resulting in more efficient internal circulation even though it is much larger. One interesting feature of Design B is the large directional skylights in the living and family rooms. These provide additional light and passive solar radiation to these large spaces and can serve as a means of ventilation in the summer. Overhangs on the upper level windows and balconies over the lower level windows allow winter sun to enter the spaces while the summer sun is screened out.

design B

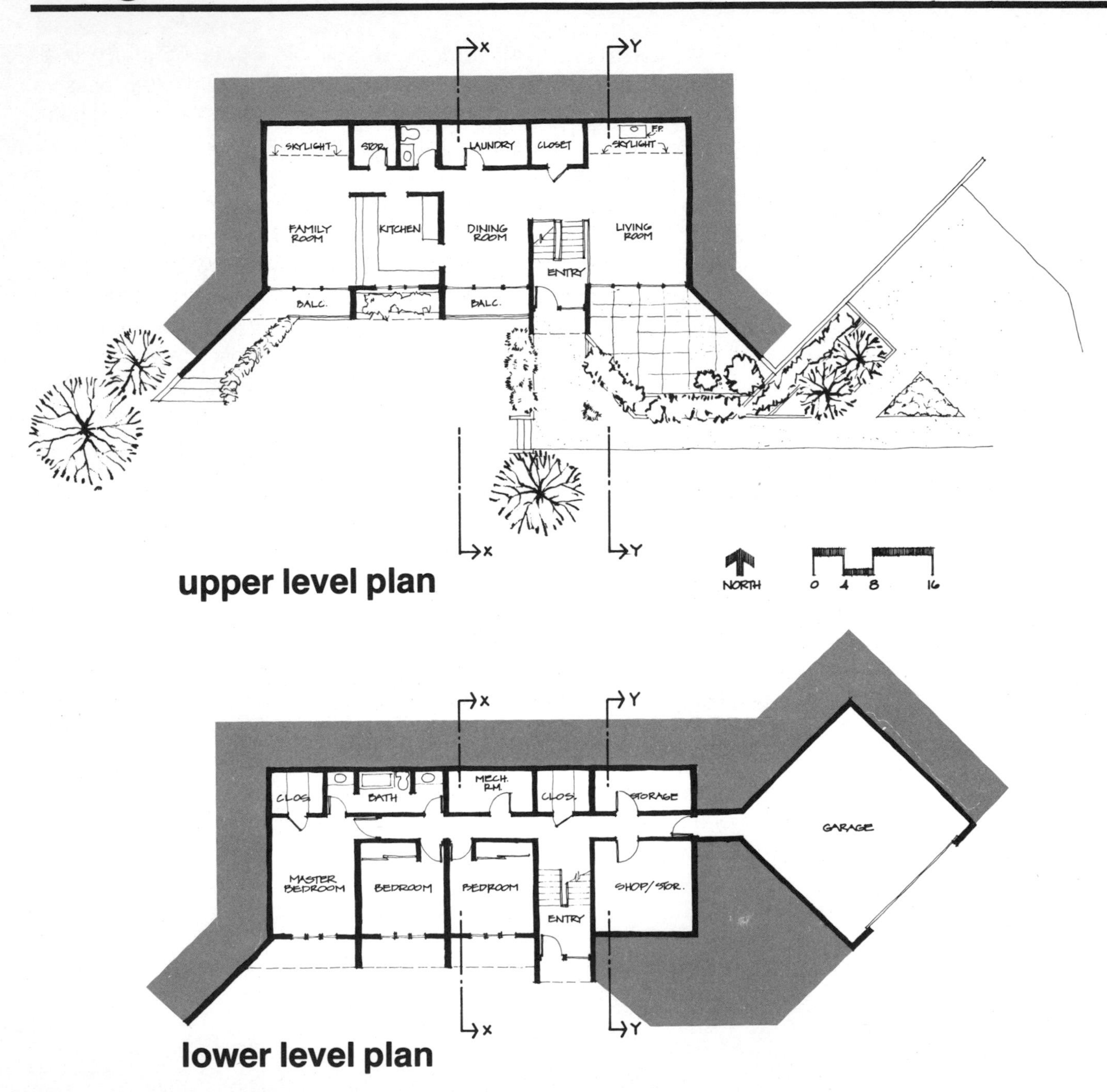

upper level plan

lower level plan

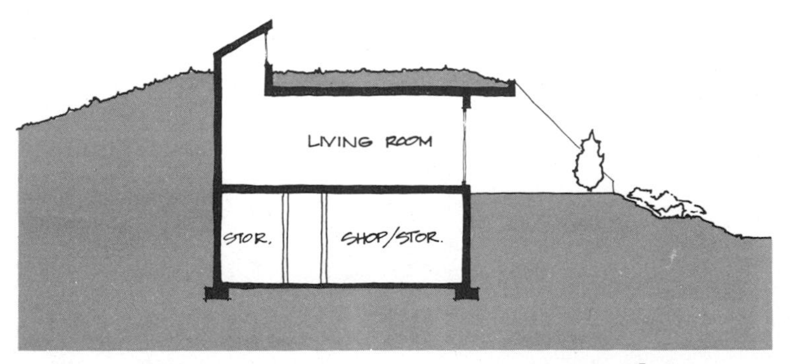

design by Carmody and Ellison

section y-y

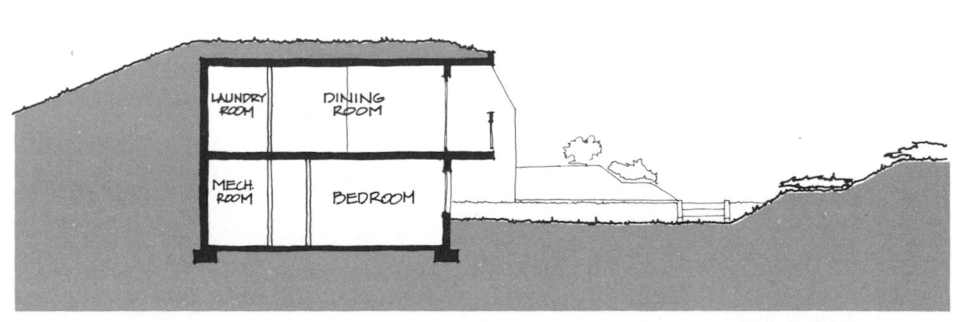

section x-x

energy use

Based on the methods of calculation used for the comparisons in Section 3, energy use projections were made for the two designs presented in this section. In each case, there are two variations of the basic design. Energy use projections for the single level house (Design A) are presented for designs with and without earth cover on the roof. One variation of the two level house (Design B) has large directional skylights over the living and family rooms, while the other has a completely earth covered roof with no penetrations. Separating the skylights in this manner illustrates their contribution to the overall energy performance of the house. The types of construction on which these calculations are based are presented in the following discussion of costs.

Although these projections are based on monthly temperature averages and are subject to a number of variables, they still represent a reasonable estimate of seasonal energy use after an initial period of three years. The results indicate remarkably low net energy requirements for both summer and winter in all cases. Several points can be illustrated by comparing the various alternatives.

The most striking result is that the two level house (Design B) actually has a lower energy requirement than the one level house (Design A) even though it has 50% more floor area. In fact, the two level house actually shows a net heat gain during the winter months, while the one level house shows relatively low heat losses for the same period. This means that the internal heat gain used in these calculations is greater than the losses due to transmission and ventilation. In addition, the net cooling load in the summer is less for Design B than for Design A even though the internal heat gain is slightly greater. There are two main reasons for the superior performance of Design B. The first is that the two level plan represents a more compact configuration than the one level plan. Although the floor area of Design B is 50% greater than Design A (2700 sq ft vs. 1800 sq ft), the total exterior surface area is only 15% greater (5770 sq ft vs. 4990 sq ft). The second reason for the superior performance of Design B is that the structure is deeper into the earth, and that it has a greater ratio of earth cover to exposed wall than Design A.

A second interesting comparison is the performance of the single level structure with an earth covered and a conventional well insulated roof (Cases A and A-2). The winter heating requirement is only slightly increased (from 1630 kW-hr to

1862 kW-hr) however, the greatest impact is in the summer cooling load. Due to the additional solar radiation reaching the roof, the total summer heat gain is nearly doubled (from 1741 kW-hr to 3315 kW-hr). It should be noted, however, that this is still a reasonably low figure when compared to most above grade structures.

design A–one level w/earth covered roof

	winter		summer
transmission	−6023		+ 41
ventilation	−1507		—
internal heat	+5900		+1700
net energy	**−1630**	kW-hr	**+1741**

design A-2–one level without earth cover

	winter		summer
transmission	−6255		+1615
ventilation	−1507		—
internal heat	+5900		+1700
net energy	**−1862**	kW-hr	**+3315**

design B–two story without skylight

	winter		summer
transmission	−3871		− 771
ventilation	−1720		—
internal heat	+6600		+1800
net energy	**+1009**	kW-hr	**+1029**

design B-2–two story with skylight

	winter		summer
transmission	−4170		− 602
ventilation	−1720		—
internal heat	+6600		+1800
net energy	**+ 710**	kW-hr	**+1198**

Finally, the energy use of the two level structure with and without the large directional skylights is worth noting. The effect of this feature is a slight increase in the winter transmission loss as well as the summer heat gain. Although, the skylights do make a noticeable impact on the energy requirements, the totals are still quite low. It should be noted that the directional skylights used in this comparison are far more energy efficient than the more typical flat or bubble units. This is because the south facing directional skylight gains passive solar heat in the winter reflecting it into the space while screening out most of the solar radiation in the summer.

Construction costs for the two designs are presented below. These costs are only preliminary and are based on a number of assumptions about the structure and the manner in which it is finished. Basically both structures are assumed to be 8 in. thick cast-in-place concrete walls with a 4 in. concrete slab floor. The roof structure and the intermediate floor in the two level plan are precast planks spanning the north-south direction thus requiring beams over the window openings to support the roof. The precast planks in Design B span in the east-west direction supported by the concrete bearing walls within the house. In Design B the south wall is a non-bearing 2 x 6 stud wall with 6 in. of fiberglass insulation. It is assumed that the walls and roof of both structures are waterproofed with a bentonite based material and that 4 in. of styrofoam insulation is placed on the roof and the upper five feet of the walls. Both designs have a forced air heating system and include a metal fireplace. All prices for windows, doors and other materials are based on good quality products and installation. The cost of Design B is given both with and without the large directional skylights over the living and family rooms. Although they have considerable aesthetic benefits, they can be considered as optional.

In addition to the two basic designs an estimate was prepared for Design A-2. This alternate design is exactly the same as Design A except that there is no earth cover on the roof. Costs for two conventional roof systems are included. One system consists of 2 x 12 rafters with plywood sheathing and asphalt shingles. This system includes 10 in. of fiberglass insulation and a sheetrock ceiling. The second system consists of laminated beams (8 ft O.C.) with 2 1/2 in. wood decking, 4 in. of styrofoam insulation and pitch and gravel roof. These two systems were selected to represent the range of costs associated with various roof systems.

Although these cost estimates are only preliminary and are subject to a number of variables, it appears that the cost of earth sheltered construction is quite comparable to good quality conventional above grade housing. Generally, construction costs for housing can vary according to location, time of year, and the amount of construction activity in the area. Constantly changing prices for basic materials have recently caused fluctuations in housing costs as well. In addition to these variables there are some unique aspects of earth sheltered construction that will also affect costs.

preliminary cost estimate

design A

1800 sq ft @ $42.27/sq ft	$ 76,100
garage	
earth covered	$ 7,325
conventional roof	$ 5,325

design A-2

conventional	
2x12 roof	
1800 sq ft @ $40.77/sq ft	$ 73,400
laminated beams	
pitch & gravel roof	
1800 sq ft @ $43.94/sq ft	$ 79,100

design B

2700 sq ft @ $40.37/sq ft	$109,000
garage (earth covered)	$ 6,855

design B-2

with directional	
skylights	
2700 sq ft @ $41.48/sq ft	$112,000

notes:
1. land costs not included.
2. earth moving and $1,000 for landscaping is included. this figures can vary up to $5,000.
3. all utilities are included except for a well (if necessary).
4. reasonable allowances for all finishes (painting, carpeting and light fixtures are included).
5. appliances are not included.
6. These costs pertain to the Mpls.-St. Paul area, spring 1978.
7. costs include 10% contractor's overhead & profit.

Contractors often will require a higher price for any innovative type of construction. This is mainly due to their unfamiliarity with certain materials and techniques. This may initially affect the cost of earth sheltered housing even though it appears to be relatively simple to construct. Once contractors have gained some experience with this type of construction and can confidently predict their own costs, the price is quite likely to go down. Another related consideration is that many typical wood frame housing contractors may not have the experience or interest to deal with a different type of structure. Many earth sheltered designs, such as the two shown here, may be more familiar to contractors with experience in commercial buildings. Thus, there may be a limited number of contractors who are capable of building this type of structure and are also interested in a relatively small project such as single family residence. This would not be a problem with a multi-unit development. It should be remembered that the construction industry is quite adaptable, and the problems mentioned here are likely to affect costs only before earth sheltered construction becomes a more familiar alternative.

In comparing the cost of earth sheltered housing to conventional above grade construction, it is essential to consider the life cycle costs of the two alternatives. The true cost of housing cannot be limited only to the purchase of land and the initial construction cost. The continuing costs of heating and cooling as well as maintenance must be included. Predicting life cycle costs is somewhat uncertain since prices change over time. However it appears inevitable that energy costs will continue to rise substantially in the future. As energy costs rise, the life cycle cost of an earth sheltered house will become increasingly favorable when compared to a conventional structure. For instance, the energy analysis section indicates that a 6000 kW-hr per year reduction in heating load for an earth sheltered house compared to a well insulated above grade house is a reasonable assumption. Making a further assumption of an energy cost (allowing for short term inflation only) of $0.06/kW-hr, the savings per year in energy costs alone will be of the order of $360 per year. If cooling costs are added this figure will become even larger.

In addition, structures with relatively little exposed exterior surface area are virtually maintenance free when compared to an above grade house. The cost of maintenance is often overlooked but can be substantial over a period of years. Even if an earth sheltered home costs slightly more to construct initially, the cost to live in it over a period of time is likely to be significantly less than in a typical above grade house. The concept of long lasting, low maintenance housing with relatively low energy requirements is very appropriate in a time of limited material and energy resources.

part b:
existing earth-sheltered houses

introduction

The most striking aspect of this collection of houses is the wide variety of shapes, details, and character of design, which clearly demonstrates the adaptibility of earth sheltered design to many tastes and needs. It can be seen that the use of earth to shelter these structures provides the medium for a variety of plant materials. This plant cover changes with each season adding a dynamic "living aspect" to the appearance of these houses.

Most of the houses use conventional structural systems, however some unique systems such as a thin shell, a culvert, and a geodesic dome are shown. An informal survey of the owners and architects have indicated that the costs for earth sheltered construction are competitive with above grade construction. According to the survey no lasting problems have been indicated and most of the owners have expressed enthusiasm for their earth sheltered homes.

This section consists of sixteen examples of existing houses and two design concepts. Since the scope of the analytical portion of this study was limited to the Minnesota region, only houses located in similar climates have been included in "Cold Climate Designs". Despite a wide range of actual conditions all other houses have been classified "Warm Climate Designs", for simplicity. It must also be noted here that the houses were not necessarily selected for energy efficiency or any special features but rather to demonstrate the wide variety of form and richness of character that earth sheltered housing can have.

A note on the drawings: the grey areas on the floor plans indicate the walls that have earth bermed against them and the grey areas on sections show the location and placement of earth.

1 cold climate designs

- winston house
- baldtop dugout
- ecology houses
- jones house
- clark-nelson house
- karsky house
- topic house
- davis cave
- big outdoors dome

LOCATION:	Lyme, New Hampshire
ARCHITECT:	Don Metz, Lyme, NH 03768
STRUCTURAL ENGINEER:	Frank Zamecnik
CONSTRUCTED:	1972
PHOTOGRAPHY:	Don Metz, Robert Perron, New York, NY

GROSS AREA:	2,800 sq. ft. including garage
EARTH COVER:	12 in.
STRUCTURE:	Concrete block walls Timber & wood deck roof
INSULATION:	Roof—1-⅝ in. compressed fiberglass Walls—zonolite in block cores Floor—1 in. urethane perimeter only
WATERPROOFING:	Roof—5 ply coal—tar pitch built up Walls—2 ply coal—tar pitch built up Floor—none
HEATING SYSTEM:	Hot water/furnace

This house was designed and built on speculation by Architect Don Metz. The house was sold before it was completed.

The Winston House is a clear example of an elevational type design which is very well suited to its mountain side site with 50 mile views to the south. The primary service and utilities areas are located along the rear wall of the house. The bedrooms and the living areas are positioned along the southern facade, with a large amount of window area to enhance the view and to maximize passive solar heat gain in the winter. The overhang on the south wall and the bearing walls help shade out the mid-summer sun. The house is constructed of concrete block masonry bearing walls (rear wall is poured concrete) with a timber and wood plank roof deck to support the 12 in. to 15 in. of earth.

The owners have expressed a great enthusiasm and have had no problems whatsoever. Architect Metz was so encouraged by the results of this project that he soon afterwards began plans for an earth covered house for himself on an adjoining site. (See the "Baldtop Dugout" on the following pages.)

winston house

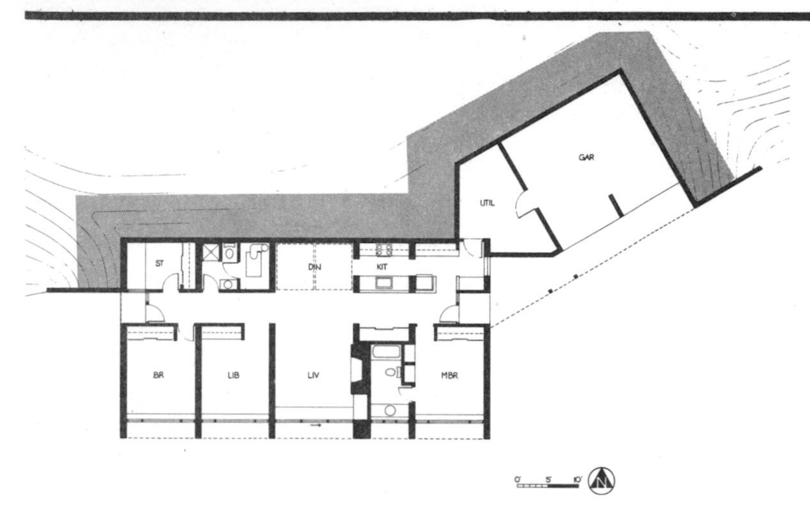

floor plan

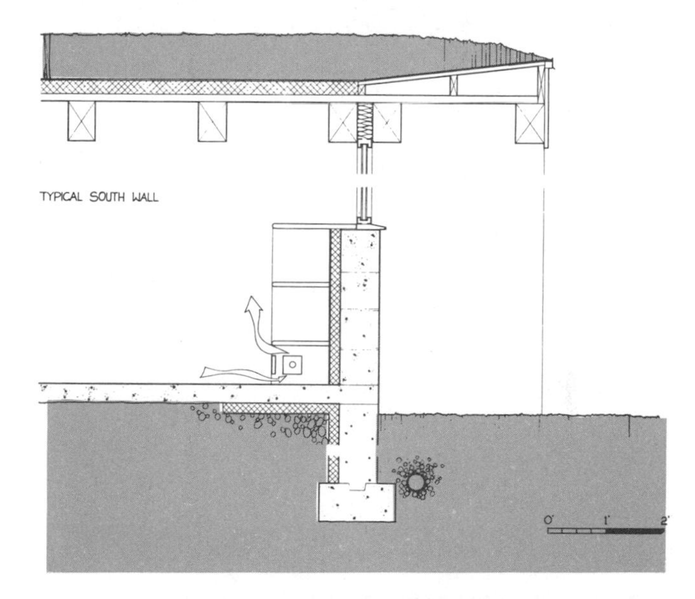

typical south wall section

baldtop dugout

Architect Don Metz's second earth covered house shares a site very similar to the Winston house, but most of the similarity ends there. Unlike the disciplined straightforwardness of the Winston house, Metz's home is characterized by curving walls, arched doorways, and much greater formality. In this house, Metz set forth to demonstrate "that building underground imposes few configurational limits, while generating a rich architectural vocabulary of its own", and that "underground building technology is simple and comprehensive, and is founded in the most traditional of building methods." Like the Winston house, Metz built this residence himself of heavy timber framing and masonry block walls.

The house is heated largely by pot-bellied wood burning stoves although an oil furnace located in the basement is available for use. The basement also provides a wood storage area that helps to humidify the house. Don indicates that eight cords of wood provided all the heat necessary for the winter of 1977. Parking is located on the upper side of the house, which is entered from the east. The earth cover has been kept to a shallow depth of 8 in., which is sufficient to support a healthy growth of native grasses and vines. Large skylights illuminate the central hallway and the work area of the kitchen.

Both designs by Metz are especially well suited to the harsh winters of New Hampshire; both also take excellent advantage of superb sites without destroying the qualities that make them desirable.

LOCATION: Lyme, New Hampshire
ARCHITECT: Don Metz, Lyme, NH 03768
CONSTRUCTED: 1977
PHOTOGRAPHY: Don Metz

GROSS AREA: 2,200 sq. ft.
EARTH COVER: 8 in.
STRUCTURE: Concrete block walls
Timber & Wood deck roof
INSULATION: Roof—3 in. urethane
Walls—2 in. urethane
WATERPROOFING: Roof—5 ply asphalt pitch
Walls—2 ply asphalt pitch
HEATING SYSTEM: Wood stoves, oil furnace back up
COOLING SYSTEM: None

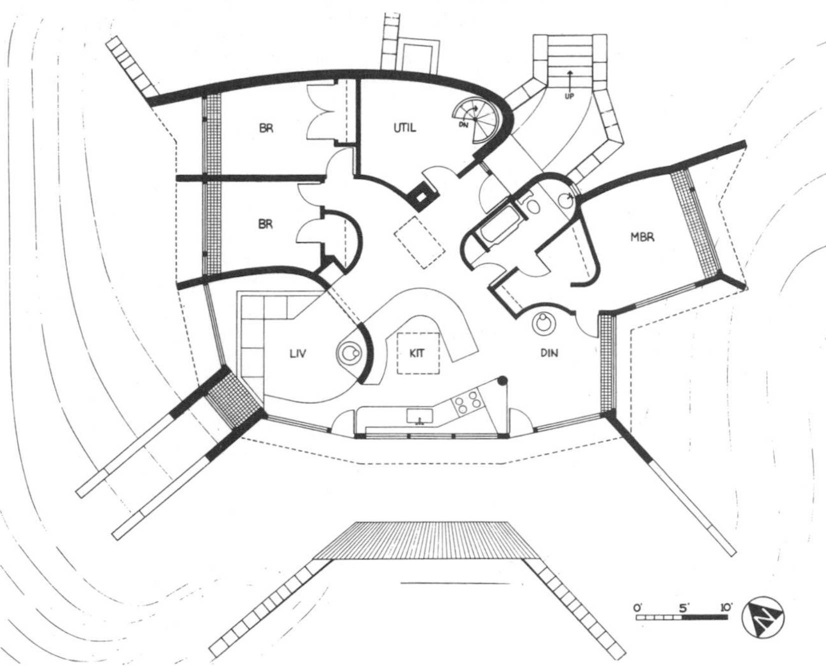

floor plan

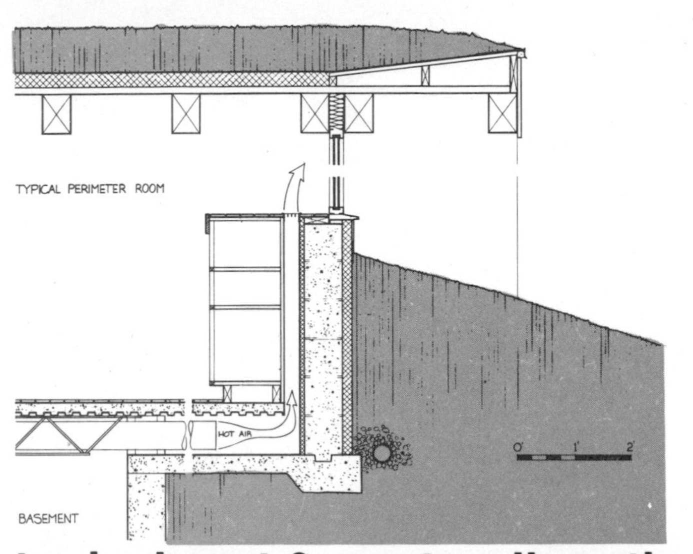

typical east & west wall section

Architect John Barnard built the first Ecology House for himself in 1973 as a vacation cottage and demonstration model to promote underground living. Tremendous publicity resulting in a great amount of interest has led Barnard to design at least four other variations on the atrium type design (one with two courtyards), plus three standard designs for sloping lot "elevational" type plans.

Barnard's style of underground design has been the model for much of the discussion and argument made in favor of subsurface dwellings. His achievement includes: a reduction of 75% in energy costs for heating and cooling, a reduction of about 25% in construction costs over conventional surface buildings, elimination of virtually all maintenance needs, fireproof construction and related lower insurance rates, superb insulation from neighboring properties and noise sources, and preservation of natural amenities—lawn, shrubs, and other greenery.

All of Barnard's Ecology House plans make use of simple, standard details that are easy to build and are certain to endure. Cast-in-place concrete walls, structural steel framing, and precast concrete plank roofing are common to all Ecology House designs. The simplicity with which the Ecology House components fit together is designed to enable the house to be built by relatively unskilled workers. This offers great potential for the "owner-builder", and could result in additional savings in construction costs.

One of the most interesting variations from the standard plans is this hillside version Barnard prepared for a family in Stow, Massachusetts. The greenhouse-like enclosure over the light court serves a passive solar collector. Air is sun-heated in the space below, and is circulated throughout the house, supplementing the regular heating and air conditioning heat pump system. The owners of this house have fully taken advantage of the earth cover on their house by growing vegetables on the roof.

Information about the Ecology House idea, and about the availability of standard or customized floor plans should be directed to John E. Barnard, Jr., 872 Main Street, Osterville, MA 02655.

LOCATION:	Osterville, Massachusetts
ARCHITECT:	John Barnard, 872 Main Street, Osterville, MA 02655
COMPLETED:	1973
GROSS AREA:	1,200 sq. ft.
EARTH COVER:	10 in. to 16 in.
STRUCTURE:	Reinforced concrete walls Precast plank roof with steel beams
INSULATION:	2 in. styrofoam on roof & walls
WATERPROOFING:	Roof—3 ply (60 lb.) built up Walls—Hot mopped pitch
HEATING SYSTEM:	Solar collector with forced air furnace
COOLING SYSTEM:	Forced air
ENERGY USE:	25% of normal

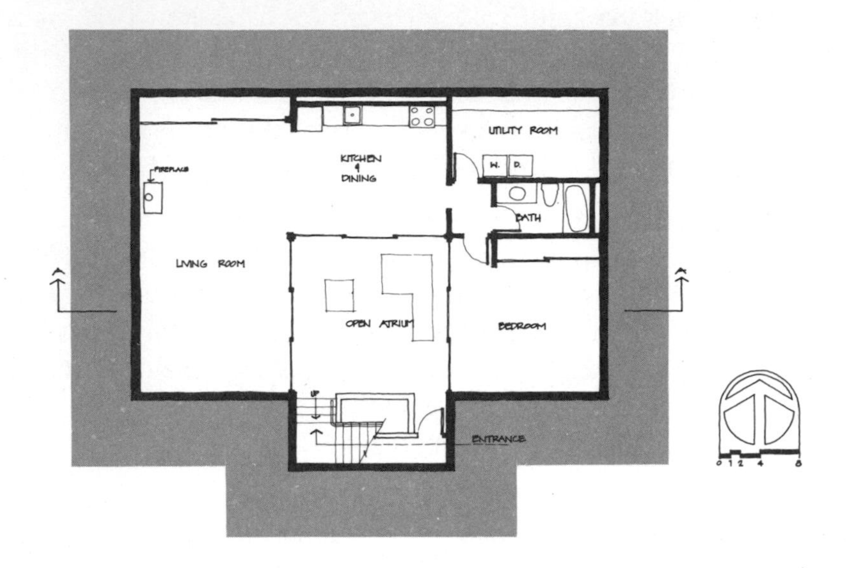

floor plan

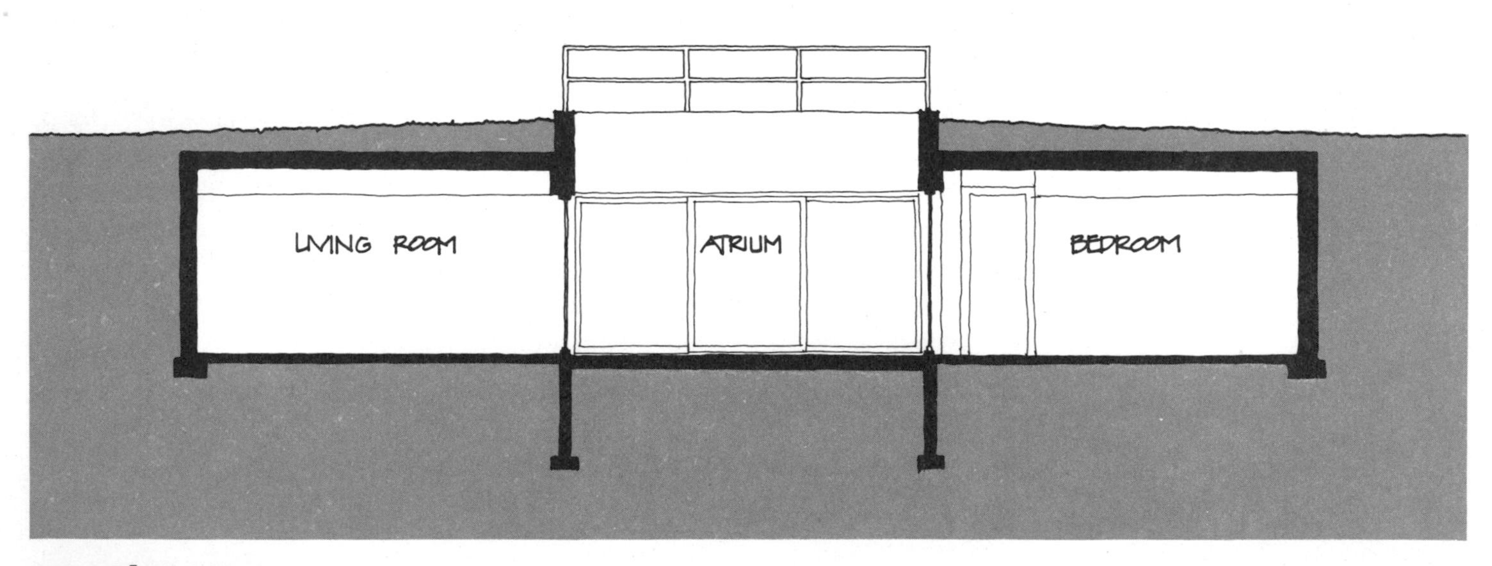

section a-a

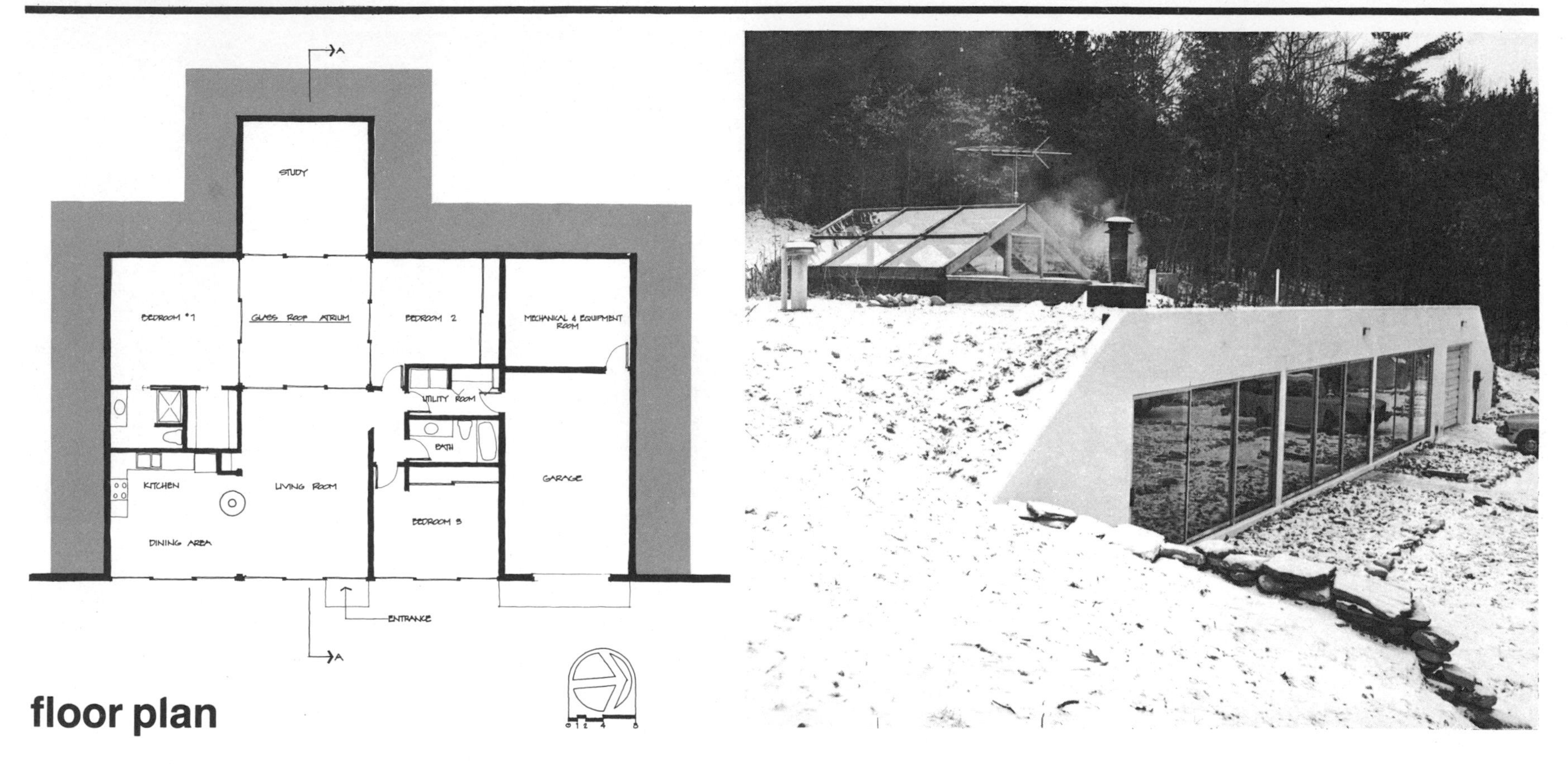

floor plan

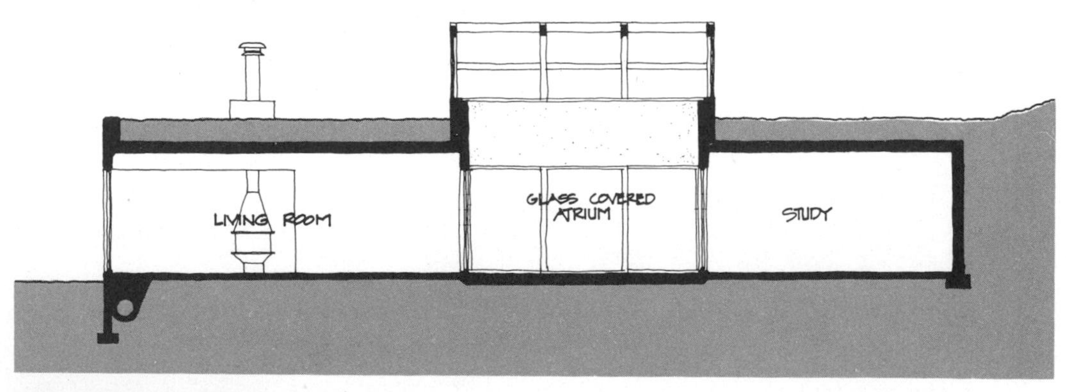

section a-a

LOCATION: Stow, Massachusetts
ARCHITECT: John Barnard

GROSS AREA: 2,300 sq. ft.
EARTH COVER: 15 in.
STRUCTURE: 8 in. reinforced concrete walls
8 in. precast concrete roof
INSULATION: 2 in. styrofoam walls & roof
WATERPROOFING: 3 ply pitch & felt roof & walls
HEATING SYSTEM: Forced air, heat pump
ENERGY USE: 50%-60% of normal house of same size

jones house

The house is approached from the east where the above grade entry and garage are the major visible forms. In addition to creating an identifiable point of entry to the house, the garage/entry structure supports a solar collector system which is oriented to the south. The structure of the house consists of poured concrete walls with a precast concrete plank roof. In contrast to most other examples, the floor is wood frame over a crawl space rather than a concrete slab on grade. The crawl space is used as a positively pressurized plenum for delivery of heat which also results in a warm floor surface. Although there is some additional cost with this type of floor system, the cost of heating ducts is eliminated. An interesting detail of the Jones house which is not attempted in any of the other examples shown here is the use of thermal breaks to isolate the interior concrete structure from the exterior concrete retaining walls and overhangs. By placing 2 in. of styrofoam between the interior and exterior elements the conduction of heat to the outside is considerably less than it would be with continuous concrete walls and roof.

Set into the top of the bluff overlooking the St. Croix River, the Jones house represents a basic approach to earth sheltered design. The house with its simple plan and single exposed elevation reflects the owner's desire to maximize the earth cover around the house and to build the structure at a cost competitive with conventional above grade construction. Although south facing windows have maximum potential for passive solar heat gain, the view of the river valley below was considered more important, consequently the exposed elevation faces west.

LOCATION: Wisconsin, Near Stillwater
ARCHITECT: Carmody & Ellison, 1800 Englewood, St. Paul
DESIGNERS: Derrick Jones & John Carmody
CONSTRUCTED: 1977
PHOTOGRAPHY: Bob Steel, St. Paul, MN
John Carmody

GROSS AREA: 1,500 sq. ft. plus 800 sq. ft. entry/garage
EARTH COVER: 12 in. (see detail)
STRUCTURE: Cast-in-place reinforced concrete walls
Precast concrete plank roof
Wood joist floor
INSULATION: Roof—8 in. styrofoam
Walls—4 in. styrofoam
WATERPROOFING: Bentonize (see roof details)
HEATING SYSTEM: Electric forced air, crawl space used as plenum, & wood stove

jones house

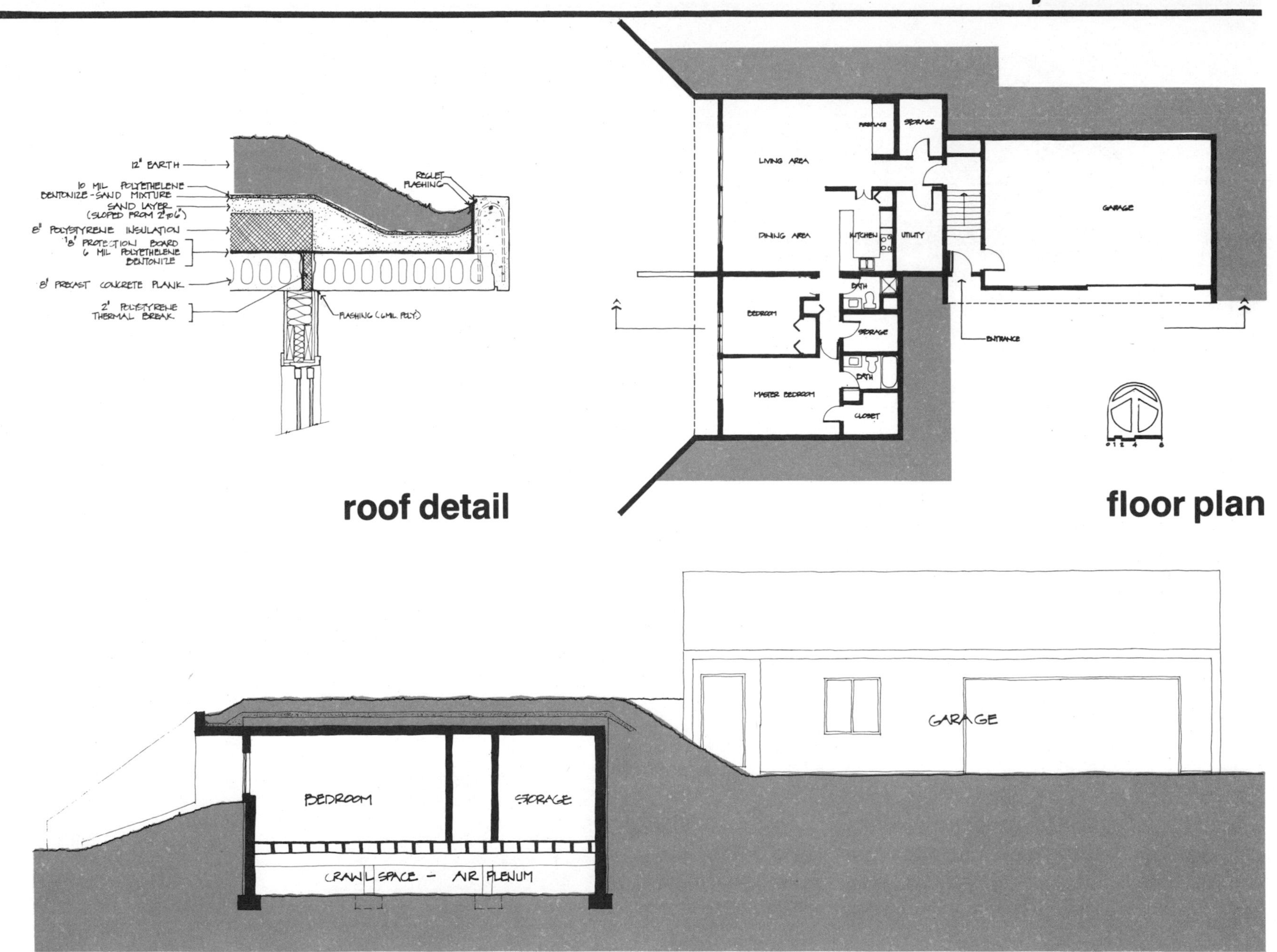

roof detail

floor plan

section a-a

Installation of styrofoam insulation and protection board over bentonite waterproofing

LOCATION:	**River Falls, Wisconsin**
ARCHITECT:	**Michael McGuire, 414 ½ S. Main St. Stillwater, MN.**
ENGINEERS:	**Paul Bredow & Assoc., St. Paul, MN**
CONSTRUCTED:	**1972**

GROSS AREA:	**2,500 sq. ft.**
EARTH COVER:	**Varies from 6 in. at peak**
STRUCTURE:	**Steel culverts w/concrete floor slabs**
INSULATION:	**2 in. polyurethane spray on interior of shell**
WATERPROOFING:	**Asphalt dampproofing and Bituthane and sheet plastic**
HEATING SYSTEM:	**Oil, forced air**
COOLING SYSTEM:	**None**
ENERGY USE:	**Reduced 25%**

The Clark-Nelson house clearly demonstrates the potential of earth sheltered housing to be well integrated into it's environment. It is set into a hillside, the arching forms complementing and blending with the surrounding shapes. Also by disturbing the existing shrubs and trees very little and by allowing the natural grasses and wildflowers to reclaim the "roof-walls", the house itself has become reclaimed by the landscape.

Taking the owners interest in new building forms, McGuire combined what he called "the advantages of a traditional sod house" with a unique use of a steel culvert structural system. The arching affect of the steel culvert was designed to economically support large earth loads, but is normally used only for drainage ditches, creeks and road work primarily due to its severe geometric restrictions. By placing two culverts side by side, one for private spaces and the other for living and common spaces and a link that houses the laundry and mechanical functions, McGuire has utilized the strict geometry to create dynamic and interesting spaces. The interior surfaces have been sprayed with polyurethane foam which provide the insulation and also give the finished plastered type appearance. However, it must be pointed out that the exposed polyurethane no longer meets the State Building Code.

In contrast with the flat roof systems of either wood or concrete which requires heavy members this shell type structure is extremely economical in its use of materials and cost. The appropriateness of this type of structure for earth sheltered construction cannot be overemphasized, it only remains to find new uses and applications of the structure such as the Clark-Nelson House.

clark-nelson house

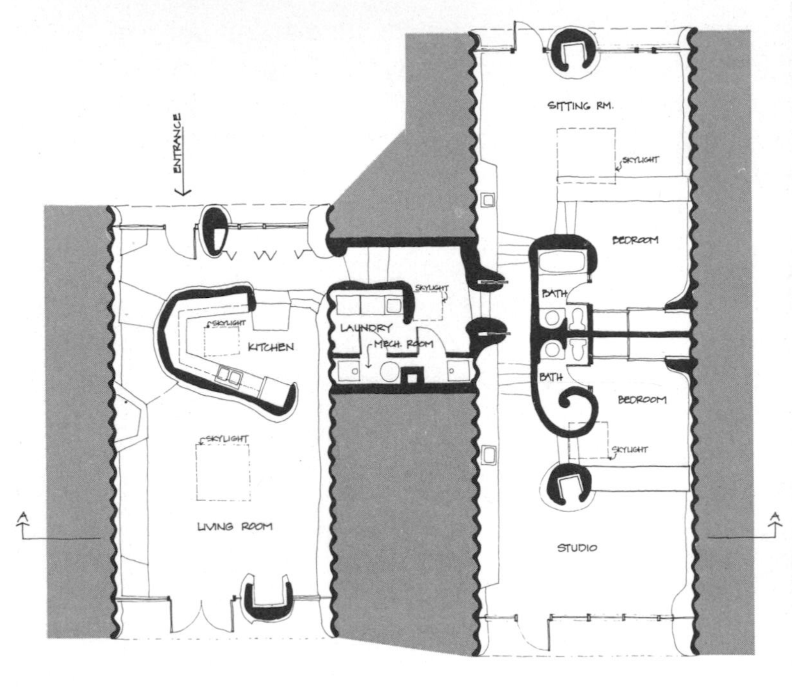

floor plan

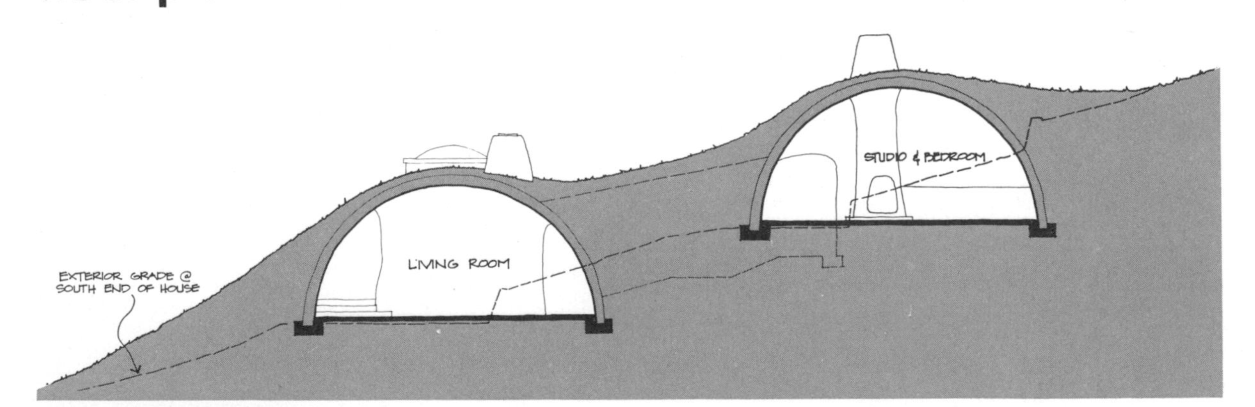

section a-a

karsky house

Mr. and Mrs. Karsky are the owners, designers and with the help of friends the builders of this house. The design and construction techniques of this house are inventive and unique. The house is set into a hill in such a way that 80% of the walls are more than 3 feet below grade level, but both levels still open to a beautiful ring shaped pond and meadow. The window wall, which faces south will be 10 in. thick and will be triple glazed. The shape of the house arose from the idea to use the same masonry blocks that are used in the construction of storage silos for farms. Interlocking flanges and a slight curvature of the silo blocks result in the circular shape of the walls, which resist earth pressure similar to the way an arch works in a wall. This facet of the house construction is very simple because no mortar or reinforcing is required, the blocks are just stacked up. (All of the blocks in the photograph below were laid in less than two days).

LOCATION: St. Croix Falls, Wisconsin
DESIGNERS & BUILDERS: Mr. and Mrs. Donald Karsky
COMPLETED: In progress

GROSS AREA: 2,000 sq. ft.
STRUCTURE: Silo block walls
cast-in-place concrete roof & floors

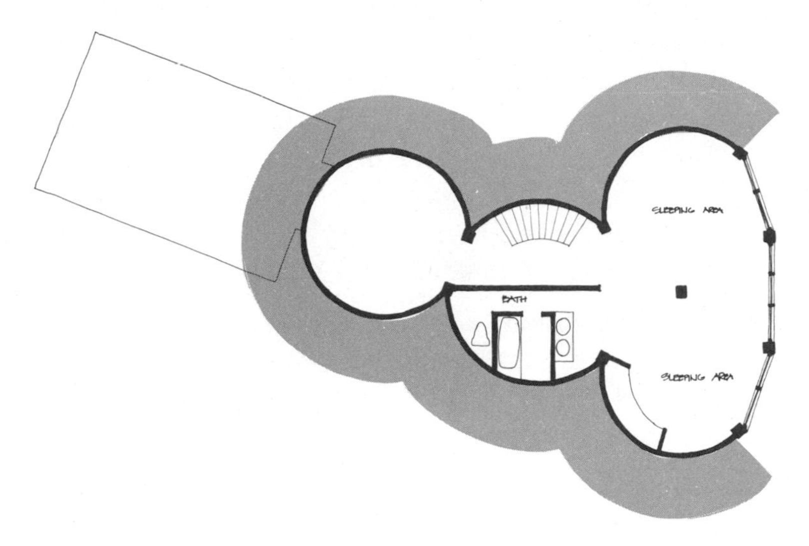

second floor plan

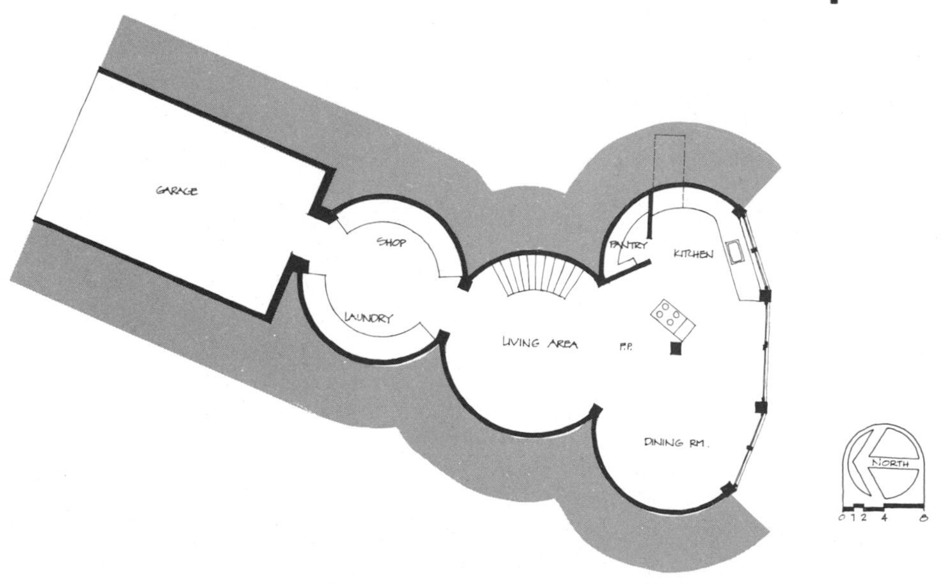

first floor plan

topic house

The Topic house is a variety of the elevational type floor plan in which all of the living spaces have windows opening on to grade level. Despite dramatic northern views over the Minnesota River valley, the house was faced south to benefit from the passive solar heating potential and to create a more private exterior space. This space is surrounded by gently sloping hills on three sides and at the center directly in front of the house a pond is being constructed. With the living areas opening up the center of the house, the bedrooms at one end and the kitchen and service areas at the other, the long "hall" character that can result with this type of plan has been minimized.

The walls of the house are precast reinforced concrete and the roof is precast hollow core concrete plank. It might be interesting to note that the exterior surfaces have been insulated with 6 in. of styrofoam on the roof and 3 in. on the walls. Also, the interior surfaces are furred out to receive 1-½ in. styrofoam insulation (walls and ceiling) and gypsum wall board. (see detail).

Since the exterior surfaces have been plastered with stucco and the windows are vinyl clad the house is virtually maintenance free. Traditionally low maintenance structures have been considerably more expensive because the materials required for such performance are more costly. However with proportionately small areas of exposed surface earth sheltered construction offers much potential to achieve low maintenance at a lower cost.

LOCATION: Shakopee, Minnesota
DESIGNER: Joe Topic
COMPLETED: 1977

GROSS AREA: 2,400 sq. ft.
EARTH COVER: 16 in.
STRUCTURE: Precast-concrete walls precast-concrete plank roof
INSULATION: Roof—6 in. styrofoam Walls—4-½ in. styrofoam
WATERPROOFING: Bentonize—roof & walls
HEATING SYSTEM: Heat pump

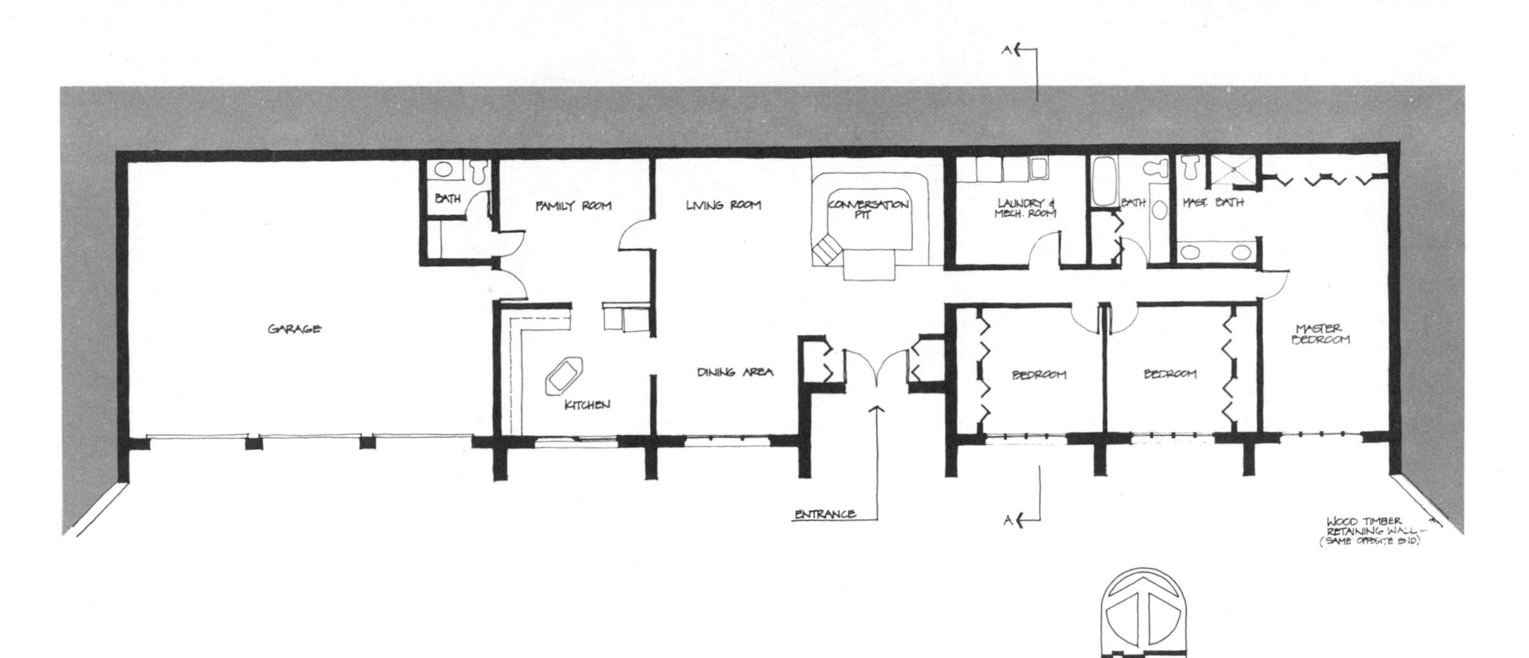

floor plan

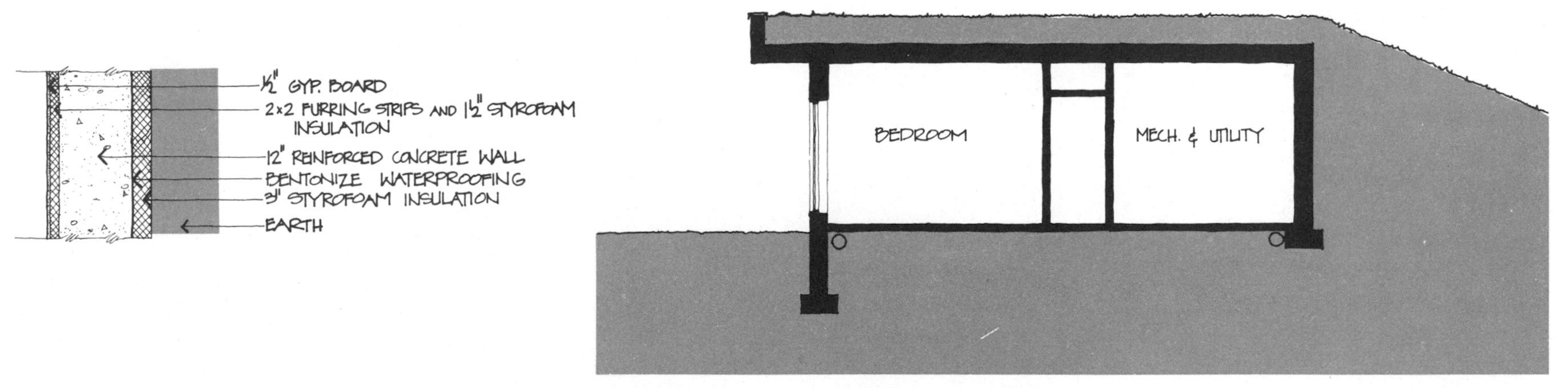

typical wall detail

section a-a

LOCATION:	**Armington, Ill.**
DESIGNED & BUILT:	**Andy Davis & Family, P.O. Box 102, Armington, IL 61721**
CONSTRUCTED:	**1975**

After seeing an abandoned mine shaft in Arkansas and being spurred on by ever increasing fuel costs for a conventional home, Andy Davis developed the idea that an earth sheltered house could be energy efficient and also a nice place to live. He had heard of a house being built this way, so he solved the problems in his own way and built the "Davis Cave" with the help of his family, for $15,000 (including the lot, carpeting and appliances).

Davis designed the original section of the house (the rectangular section was added for his mother-in-law) as an octagon to limit the length of unsupported walls. The entire shell is made of cast-in-place reinforced concrete; the roof is 12 in. thick, the back and side walls are 8 in. thick and the front and only exposed wall is 12 in. thick. When asked about the rocks that cover the interior surfaces, Davis replied "The most important reason the rocks are there is because they really look nice and they give this house the atmosphere we were after. We wanted to live in a cave . . . and the stones all over the walls and the ceiling really do make our home feel like a cave. The second reason we covered the inside of our house with rocks is because they should stay just as good looking as they are now dang near forever."

The first winter the home was heated solely by burning 2 & ½ cords of wood in a Franklin fireplace. At one point the Davis family was snowed in by a blizzard for four full days with no wood to burn. After an attempt to cut up tires for fuel, they let the fire go out. Davis described what happened: "We only lost two degrees a day. It was something like 25° below zero outside with a wind chill factor way down to a minus 80°, but in four days without a fire, the temperature in here only slid down from 70° to 62°. To tell the truth, I could have gotten out and brought wood in a day or two earlier than I finally did . . . but we were all so comfortable that I didn't want to do it. We just didn't have anything to worry about."

Presently Andy Davis is selling the franchise rights to this type of house, for further information contact him through the address at the left.

GROSS AREA:	**1,200 sq. ft. plus 800 sq. ft.**
EARTH COVER:	**3 ft. to 4 ft.**
STRUCTURE:	**Roof, walls & floor—cast-in-place, reinforced concrete**
INSULATION:	**Roof—1 in. styrofoam Walls—1 in. styrofoam extending down 8 ft. from surface**
WATERPROOFING:	**Walls & roof—asphalt coating**
HEATING SYSTEM:	**Franklin fireplace**
COOLING:	**None required**

davis cave

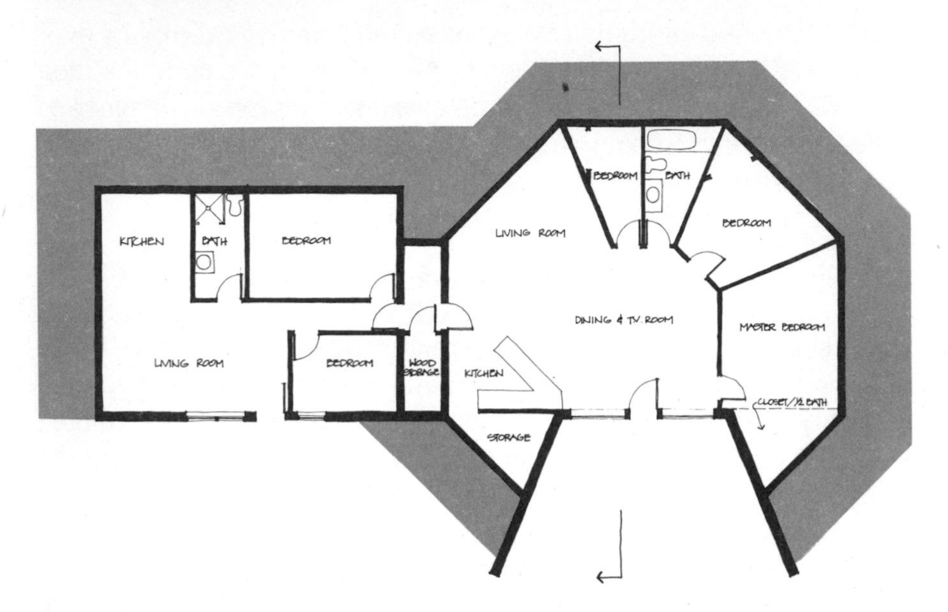

floor plan

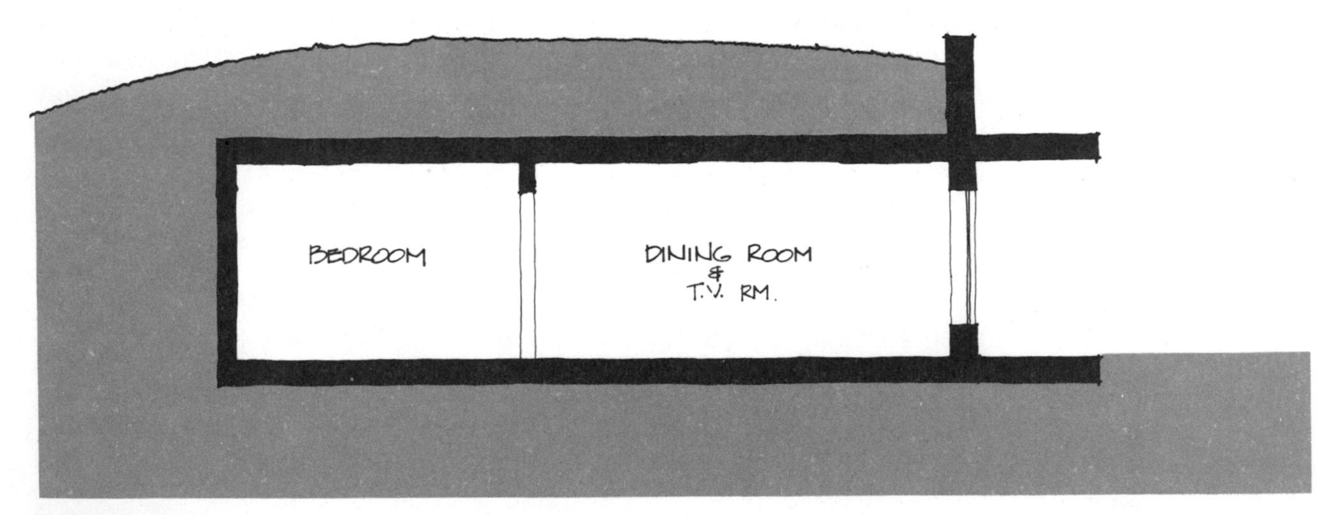

section

big outdoors dome

This design concept links together 3 geodesic domes and sets them into a south facing slope sheltering the rear of the house with earth. One of the larger domes provides privacy for three bedrooms and baths while the other has a living area loft above and open to the kitchen, eating and family areas below. The smaller center dome is a screened porch which also acts as a canopy for the entry. The living space dome has a window cluster arranged in the form of an "Iris" and the bedroom dome has a solarium. Both window openings have shading devices to minimize summer solar heat gain and shuttering devices to maximize winter solar heat gain. One advantage of the geodesic dome is that it can enclose more space per exterior surface area than any other shape. This unique structural system, which uses relatively little material to support large loads, appears to be well suited to earth sheltered housing.

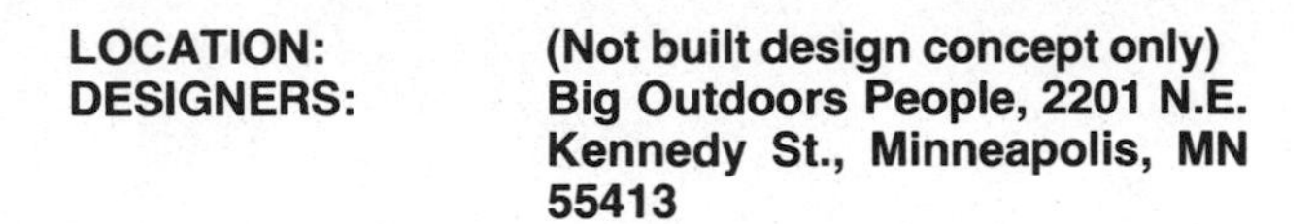

LOCATION: (Not built design concept only)
DESIGNERS: Big Outdoors People, 2201 N.E. Kennedy St., Minneapolis, MN 55413

GROSS AREA:
EARTH COVER: varies
STRUCTURE: 4 frequency, 30 ft., Geodesic Dome, with ½ truncation
INSULATION: 7-½ in. urethane foam
HEATING SYSTEM: Forced air

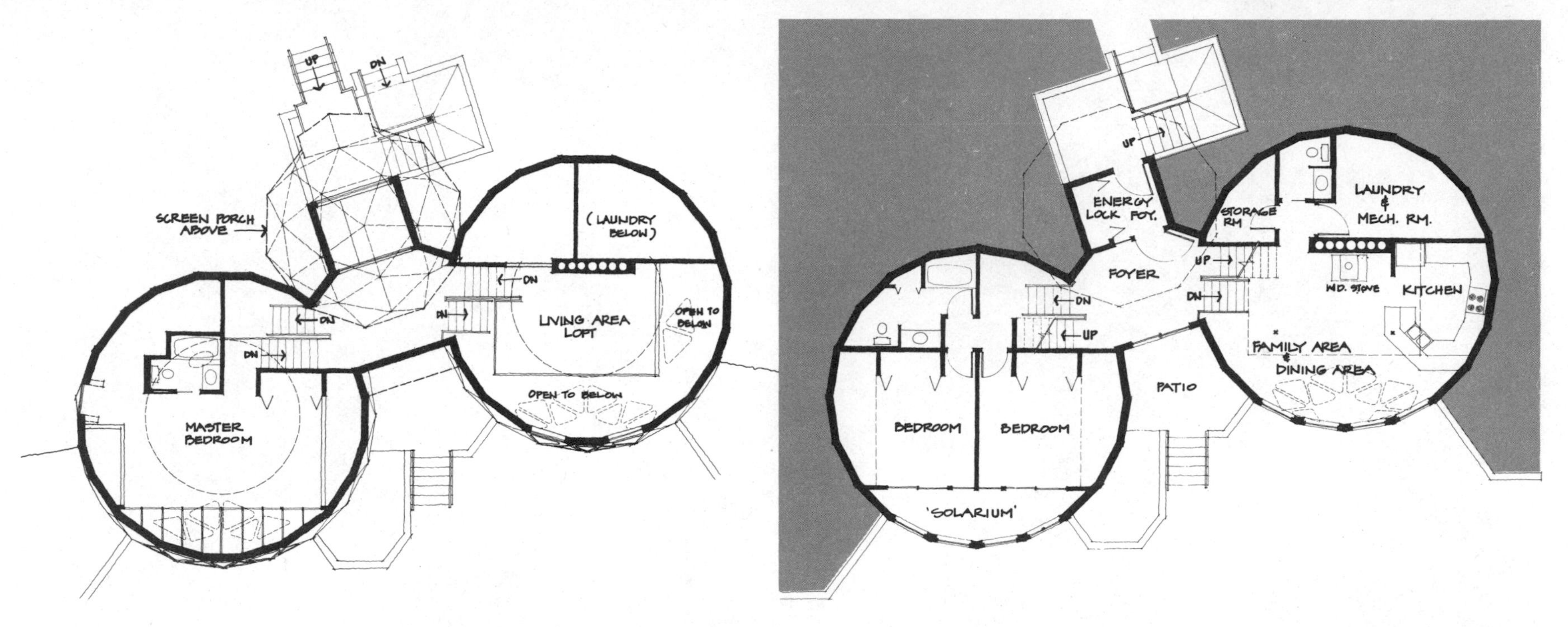

second floor plan

first floor plan

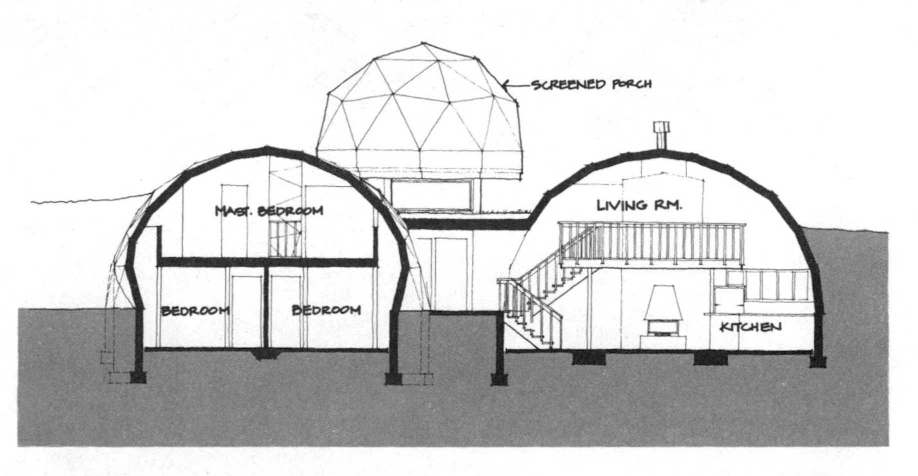

section

2 warm climate designs

- solaria
- rousselot house
- bordie residence
- dune house
- sundown house
- alexander house
- geier house
- terrasol

LOCATION:	**Vincentown, New Jersey**
ARCHITECT:	**Malcolm Wells, P.O. Box 1149, Brewster, MA. 02631**
CONSTRUCTED:	**1975**

GROSS AREA:	**2,800 sq. ft.**
EARTH COVER:	**24 in. (12 in. sub soil) (8 in. mulch) (4 in. top soil)**
STRUCTURE:	**Concrete block walls Timber & wood deck roof**
INSULATION:	**2-½ in. styrofoam—roof, walls & floor edge**
WATERPROOFING:	**Roof—⅛ in. butyl rubber Walls—Foundation coating on parged concrete block**
HEATING SYSTEM:	**Solar collectors (Thomason type) with oil back up**
COOLING SYSTEM:	**Conventional a/c with rock storage**

In contrast with some of the designs presented in this report Solaria represents a serious effort to incorporate energy efficiency with the aesthetic aspect of earth sheltering. It has been described as the first solar-heated, earth covered house in New Jersey. It is not truly "subsurface" in that it's sod roofcover is discontinuous with all exterior grades. The structure is earth sheltered because the roof is generously blanketed with up to two feet of earth and mulch. Lifting the roofline above the exterior grade on the north side allows natural lighting at the rear, aids cross ventilation and provides entry without making a major change of grades.

Architect Malcolm Wells plan for Solaria is organized much like an elevational scheme—all major living areas have generous southern exposures, and most all utility and non-living spaces are located along the north wall, with its smaller window areas. The structure is of heavy timber framing, with a wood plank roof. The tall southern facade makes plenty of room available for full height windows from floor level, plus an entire array of Thomason solar collector panels. The height at the front also opens up large interior spaces with adequate room for second level lofts.

Wells reports in his booklet UNDERGROUND DESIGNS that the earth covered, solar-heated combination worked so well that in the winter of 1977, no auxiliary heat was required to supplement the Solaria heating system. A documentary book describing the entire construction process, and the design and details of Solaria is available from the Edmund Scientific Corporation, Barrington, NJ. Also available from Edmund is a license to build Solaria, a set of slides depicting the house and its components under construction, and information regarding Thomason solar heating systems.

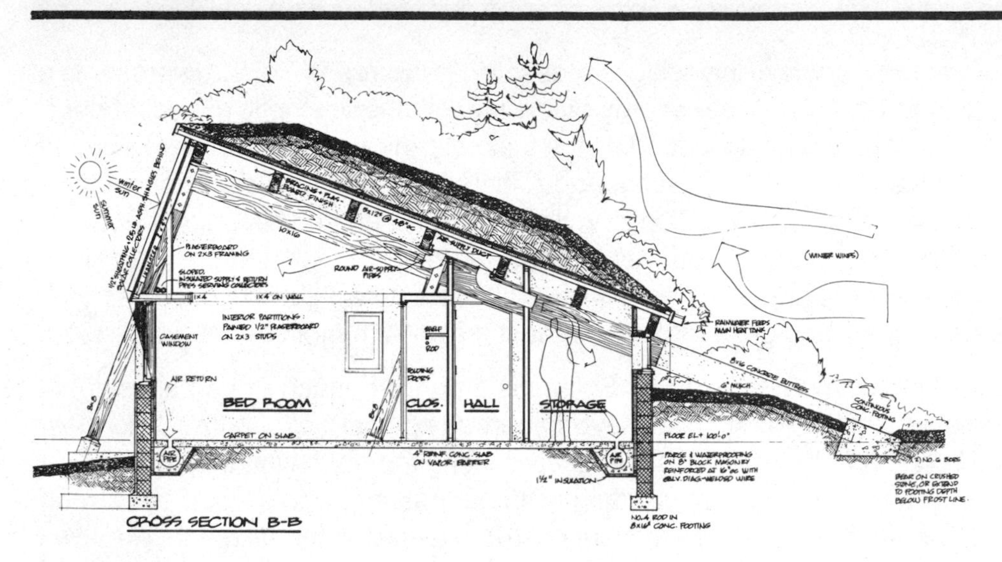

section b-b

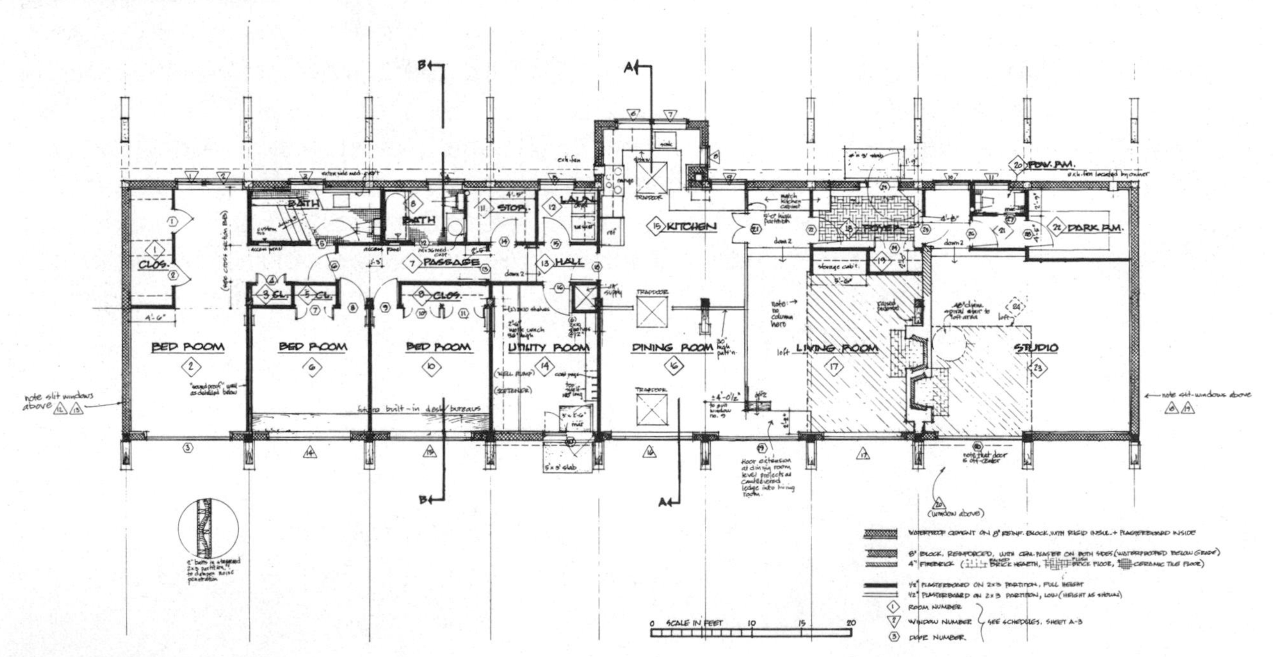

floor plan

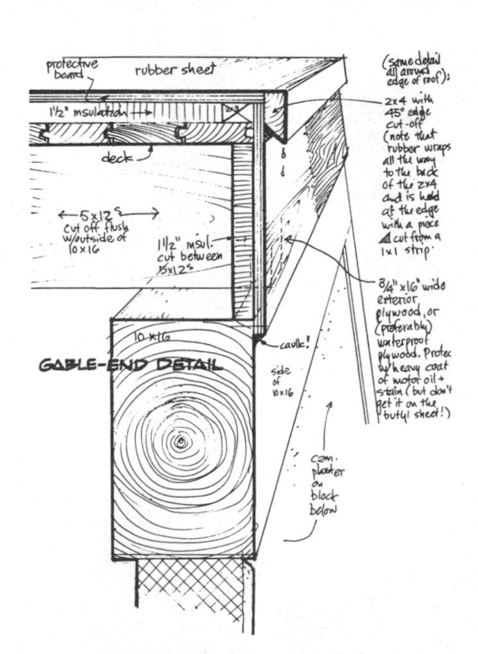

gable end detail

rousselot house

This low profile dwelling is a sensitive and sensible interpretation of a style and of images native to the American Southwest. The house is distinctly of New Mexico in character, due to its rounded corners and detailing, and its adobe-like use of masonry bearing walls. The design intentionally alludes to the historic Indian pit houses indigenous to the area, and to the sod house that once sheltered the owner's pioneer grandmother. These historical allusions are more than symbolic or superficial, since the earth cover and heavy construction of this modern dwelling perform the same thermal climatic functions as did the corresponding features of the pit and sod houses.

Architect Ron McClure welcomed this opportunity to practice an interest in underground design that he has maintained since leaving college. He views earth covered construction as a means to energy conservation through climate responsiveness, and as a way of integrating architecture with the natural landscape. This particular design is intriguing in that each side of the house differs considerably from the others in appearance, giving the impression that the structure is very complex in its organization, in fact, the plan is quite simple and straightforward. Complementing the clarity of internal order is a richness of detail and molded angular forms that express the supporting structure of the roof. Aptly, this residence has been described as a "paradox house", a contradiction of being underground, but having views of the majestic Taos landscape from all sides.

LOCATION: Taos, New Mexico
ARCHITECT: Ron McClure, Box 628, Tijeras, N.M., 87059
ENGINEER: Buck Rogers, Rininconada, N.M.
CONSTRUCTED: 1971

GROSS AREA: 3500 sq. ft.
EARTH COVER: 12 in. to 16 in.
STRUCTURE: Roof—4 x 6 tongue & groove wood deck
Concrete block walls
INSULATION: Roof—sod only
Walls—pumice in blocks
Floor—batt insulation
WATERPROOFING: Built up roofing
Bituminous coating on walls
HEATING SYSTEM: Electric, forced air

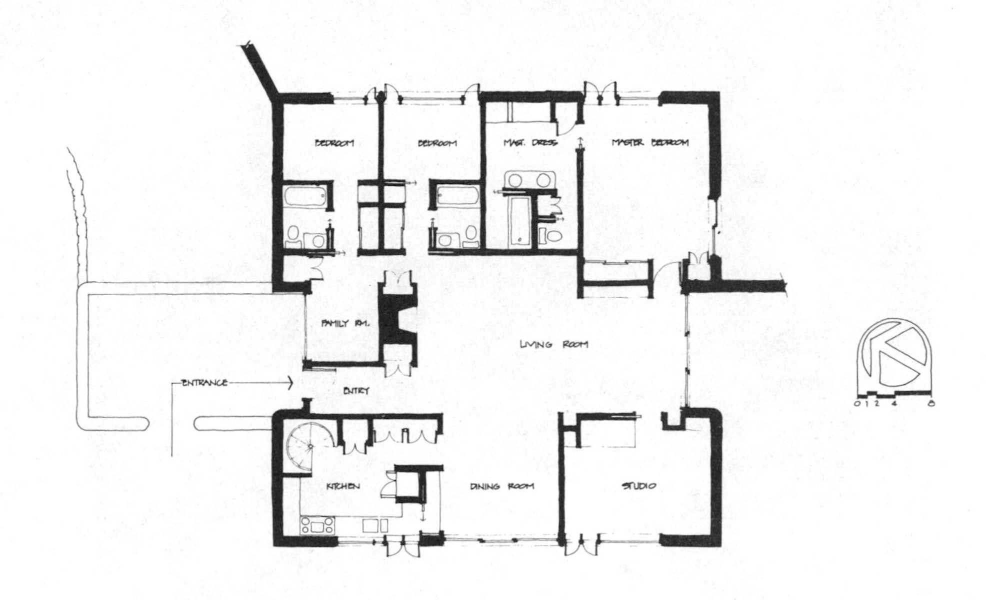

floor plan

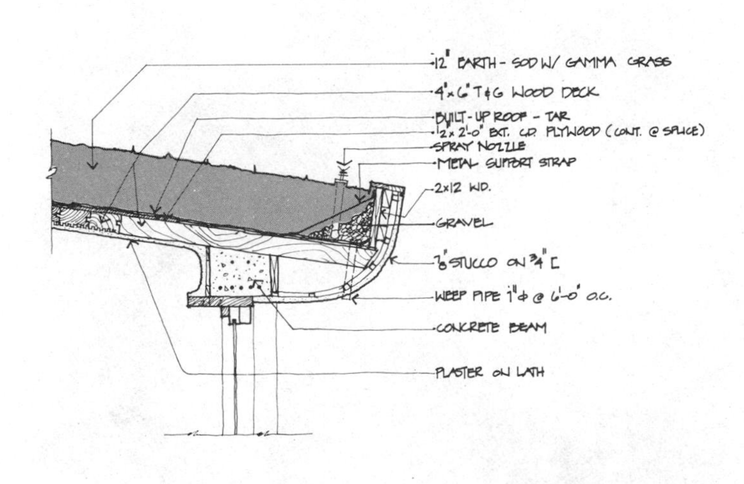

roof detail

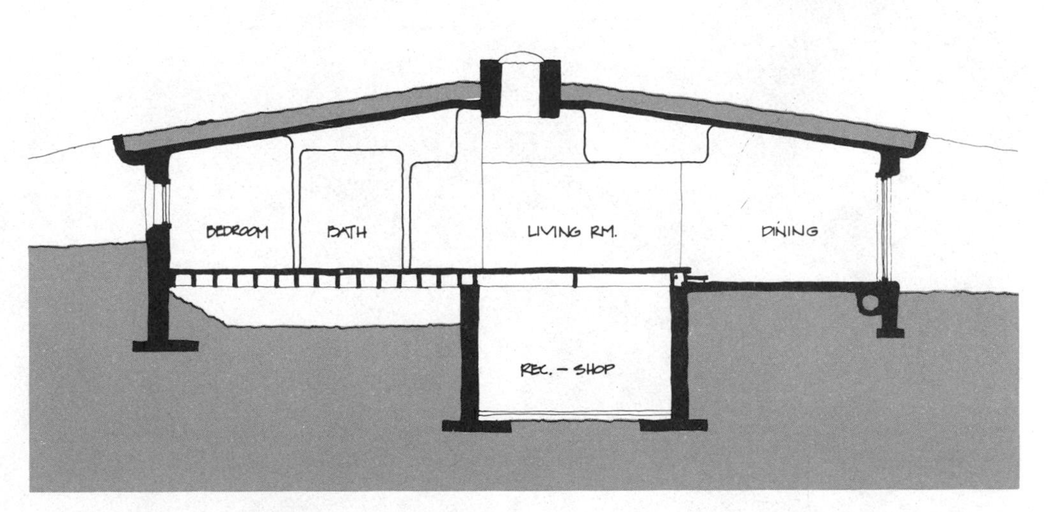

section

LOCATION:	Austin, Texas
ARCHITECT:	Coffee & Crier, 509 Oakdale, Austin, TX.
ENGINEER:	George Maxwell
CONSTRUCTED:	1975
PHOTOGRAPHY:	Coffee and Crier

GROSS AREA:	2,000 sq. ft.
EARTH COVER:	14 in. to 18 in.
STRUCTURE:	8 in. reinforced concrete walls Precast concrete roof
INSULATION:	Roof—3 in. urethane foam
WATERPROOFING:	Roof—sheet membrane Walls—mastic & fiberglas, mopped on Under Floor—plastic sheet
HEATING SYSTEM:	Solar Hot water w/electric furnace forced air back up
COOLING SYSTEM:	Direct expansion split system, electric powered, 2 ton
ENERGY USE:	About 50% of typical construction

A multitude of conditions imposed by both the climate and the owner were satisfied by this subsurface design. In the face of increasing fuel and energy costs, the client wished to build a house that would be comfortable, yet cost little to maintain. The Texas location makes solar heating a particularly attractive alternative, and placing the structure underground makes the solar heating even more practical: a 50% reduction in utility bills has already been realized. The recent history of two tornadoes passing through this property presented another very real concern to the owner, who expressed a personal desire for an atrium type plan. This type of an arrangement, with a fully enclosed "exterior" court is extremely well suited to earth sheltered design.

The approximately 20 ft. x 30 ft. courtyard is an integral part of the plan in many ways. In addition to admitting daylight and air into surrounding rooms, and serving as an outdoor room in itself, the recessed atrium is actually the central circulation space, connecting the childrens' bedrooms to the main living areas of the house. This exterior detachment is purposeful in ensuring privacy between the different "wings" of the house, and will serve this same function when the children have left and their rooms are converted into studies or offices. Although this exterior circulation arrangement may only be satisfactory in southern climates, such a plan could readily be roofed with a demountable cover or inflatable dome for protection from northern winters.

The Bordie House is not fully subgrade, in fact, the house has been only halfway recessed into the top of a low knoll and mounded over with 14 in.-18 in. of earth. The slope of the underground concrete plank roof gives height to the rooms below, and accentuates the mounded form above. Solar collector panels are mounted on a similarly mounded form independent of the house, and all earth cover on and around the house has been replanted with native grasses and wildflowers.

bordie residence

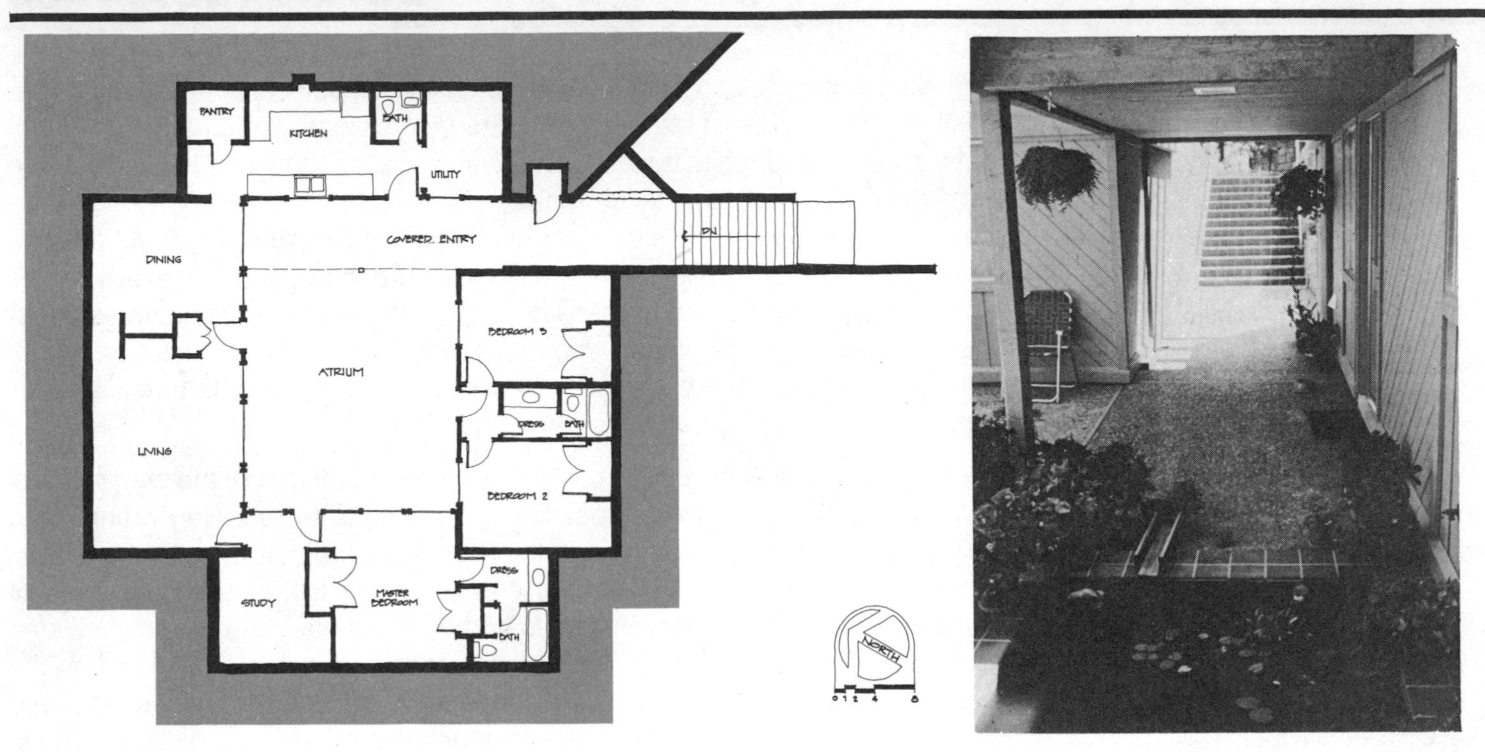

floor plan

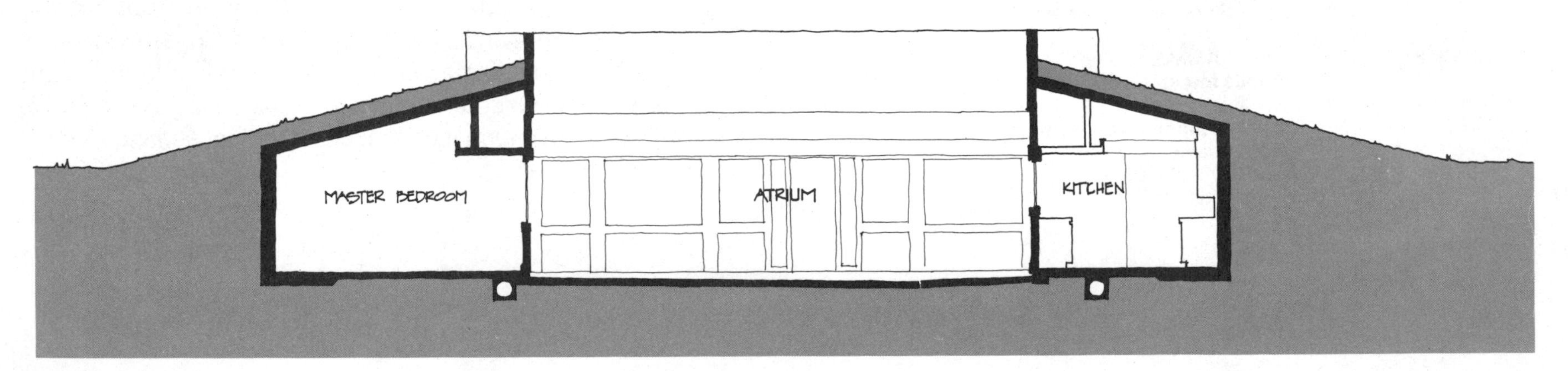

section

LOCATION:	Atlantic Beach, Florida
ARCHITECT:	William Morgan, Architects 220 E. Forsyth St., Jacksonville, Fl. 32202 William Morris, Project Architect
STRUCTURAL ENGINEER:	Geiger Berger Assoc.
MECHANICAL ENGINEER:	Roy Turknett Engineers
CONSTRUCTED:	1974
PHOTOGRAPHY:	Alexandre Georges Creative Photographic Services

GROSS AREA:	750 Sq. ft. each unit (1,500 total)
EARTH COVER:	22 in. minimum
STRUCTURE:	Reinforced gunite shell
INSULATION:	None required
WATERPROOFING:	Liquid bituminous, brushed on
HEATING AND COOLING:	1-½ ton water cooled reverse cycle heat pump

William Morgan has been designing earth related structures for many years, and has recently completed several houses which are bermed around the exterior walls, but roofed with conventional materials. The Dune Houses shown here are Morgan's first earth-roofed dwellings, and are owned and maintained by Morgan Properties as rental units. They are unique in many respects, not least of which is the fact that they are the first (modern) underground multi-family units to be built in this country. The Dune Houses are also the first subsurface dwellings to fully exploit a structural system designed expressly in response to the peculiarities of earth pressure loading.

Energy conservation was not a prime consideration in the design of these units, since the Jacksonville area is blessed with a mild climate all year long. Still, the subsurface design maintains an almost isothermal 70° indoor temperature, although this is much affected by the residents habits regarding cooking, lighting, washing and drying, and leaving doors open. Morgan's main objective in building below grade was to preserve the low profile of the beach front property from the street side.

Morgan described that "one of the main ideas behind this design has been designing a building that is of the earth, not just in the earth—i.e., an above grade design uncomfortably overstressed when below grade. We designed the shells in careful balance with the surrounding earth so that the inward pressure of the earth presses uniformly on the shell, locking or post tensioning the gunnite shell in place. The structure actually is stronger because of the earth pressing on it." Though the use of shell type structures Morgan's approach to design may prove to be one of the major directions for earth sheltered housing.

dune house

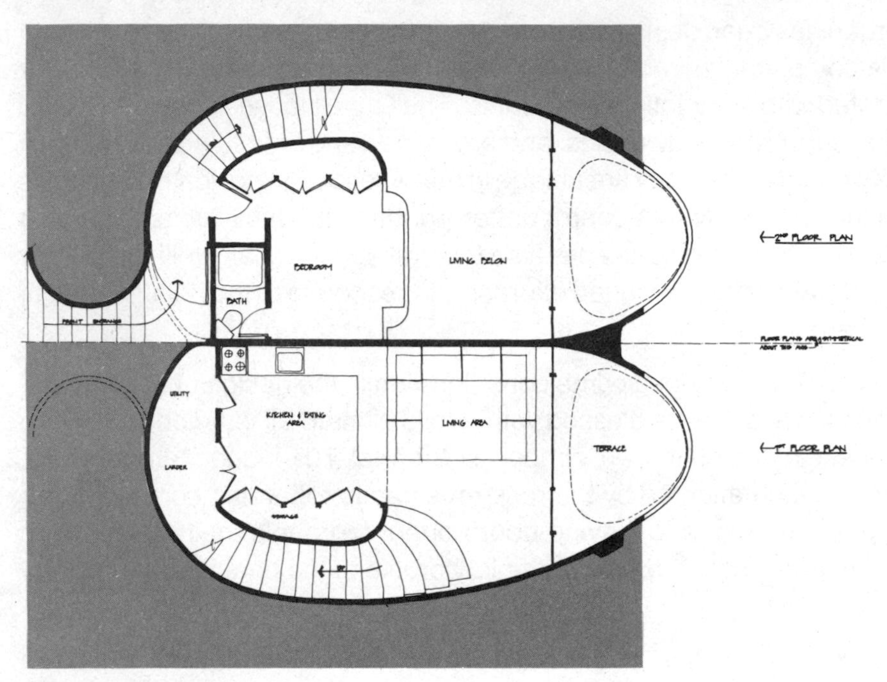

floor plan

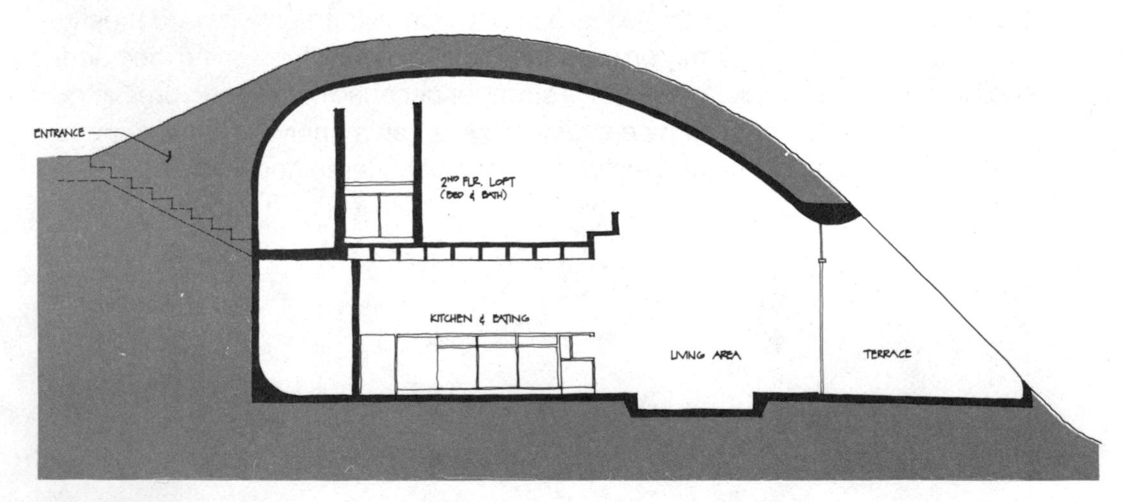

section

sundown house

Architect David Wright, has been designing solar-heated houses in the Southwest for several years, so it was a logical decision for him to take his quest for an environmentally sensitive architecture one step further, and build his own solar-heated residence underground. The continuous grade of the sod roof makes Sundown invisible from the shore of his site along the California coast. Native meadow vegetation has reclaimed the roof, blanketing the house in wild grasses that shade the soil from the winter sun.

There is no lack of sunlight inside the house, since it has been oriented to optimize penetration of winter solar radiation. The roof is pitched rather steeply to accommodate the low winter sun, which provided 99% of the heating requirement for the winter of 1976-1977; ten pieces of oak provided the remaining hot water needs. The passive solar design makes use of finish and structural materials, such as the rear concrete wall and brick flooring, to store the sun's energy. Operable roof skylights admit daylight to the recesses of the interior, and are hinged to vent rising hot air in the summer. It is interesting to note that the exterior "atrium like" space has been created by careful placement of the garage/studio and earth berms in relation to the house and by being partially recessed. This creates a "weather-conditioned" front yard that is fully exposed to the south, but sheltered from the prevailing winds. Wright believes that solar energy and subsurface design are ideal complements for environmental architecture, and is currently designing several houses for different sites throughout the Southwest embracing these principles.

LOCATION: Sea Ranch, California
ARCHITECT: David Wright, Environmental Architect, P.O. Box 49, Sea Ranch, CA 95497
CONSTRUCTED: 1976
PHOTOGRAPHY: Owen Brewer, Darrow M. Watt, Portola Valley, CA

GROSS AREA: 1,200 sq. ft. plus 400 sq. ft.
EARTH COVER: 6 in. w/native grass
STRUCTURE: Reinforced concrete block walls
Wood rafters & plywood roof
Brick on sand floor
INSULATION: Roof—2 in. styrofoam
Walls—2 in. styrofoam
Floor—1 in. styrofoam
WATERPROOFING: See Detail
HEATING SYSTEM: 95% passive solar
COOLING SYSTEM: Naturally induced ventilation (see section)
ENERGY USE: 1% to 5% of normal

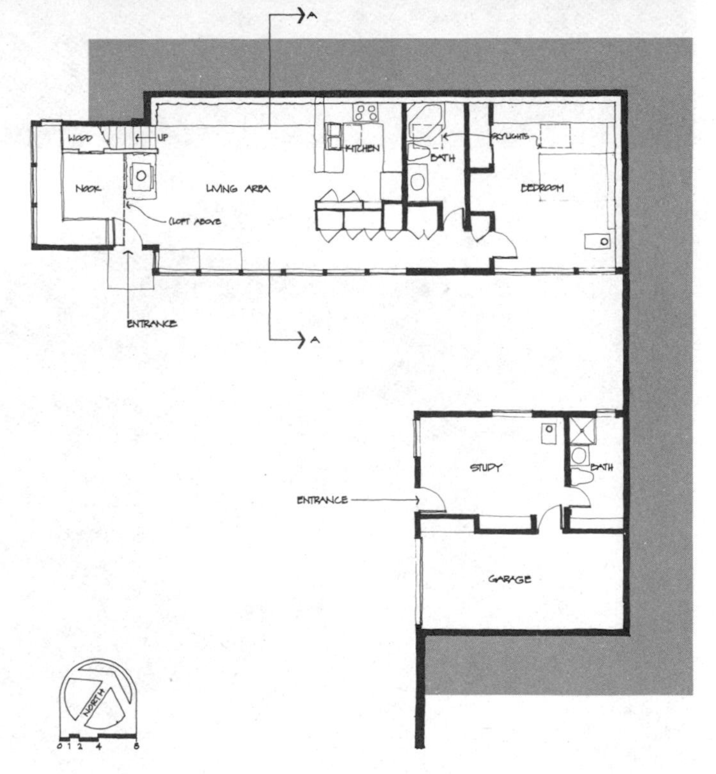

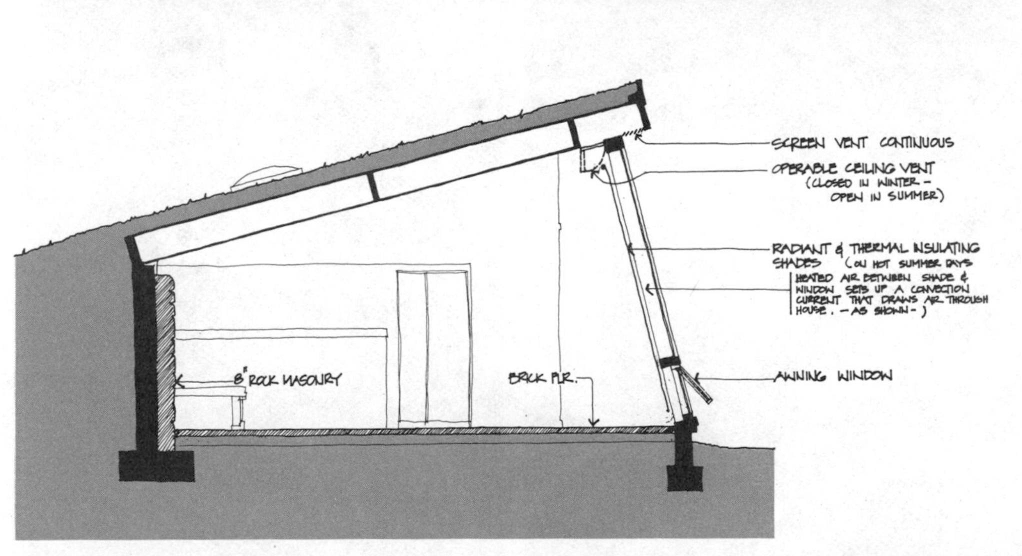

section a-a

floor plan

sundown house

alexander house

LOCATION:	**Montecito, California**
ARCHITECT:	**Roland Coate, 1905 Lincoln Bl., Venice CA. 90241**
CONSTRUCTED:	**1974**
PHOTOGRAPHY:	**Jurgen Hilmer, Santa Barbara, CA**

GROSS AREA:	**7,000 sq. ft.**
EARTH COVER:	**varies**
STRUCTURE:	**Reinforced cast-in-place concrete for walls, roof, floor**
HEATING SYSTEM:	**Gas; fired, forced air**
COOLING SYSTEM:	**Installed but not needed**

Unquestionably one of the most imposing residences built anywhere in recent years, this house commands a magnificent view over the Pacific Ocean in Montecito, California. The susceptibility of the region to brush fires—provides a basis for this design, but most directly the integration of the house with earth reflects an interest that architect Roland Coate had developed over a period of years and was eager to explore. Credit must also be given to the owner, whose own imagination and support have helped make this house "one of the major pieces of serious residential architecture to have emerged in California in recent years" (Paul Goldberger, architecture critic, NEW YORK TIMES).

This project is really not one, but two houses. The principal residence is a two level structure which contains major living areas plus a studio, study and darkroom. Window exposures with views are provided on three of four sides, and a recessed courtyard provides natural lighting for the fourth interior wall of the main living level. This atrium area serves as the main entry court, as well as a pleasant outdoor patio dining area. A circular stair tower connects all vertical levels, and creates a gazebo-like lookout on the rooftop. A detached guest cottage (not shown here) upslope from the main house is designed somewhat in the form of a dugout, with bedrooms facing the east, and with a corresponding dugout area containing parking garages facing the west.

alexander house

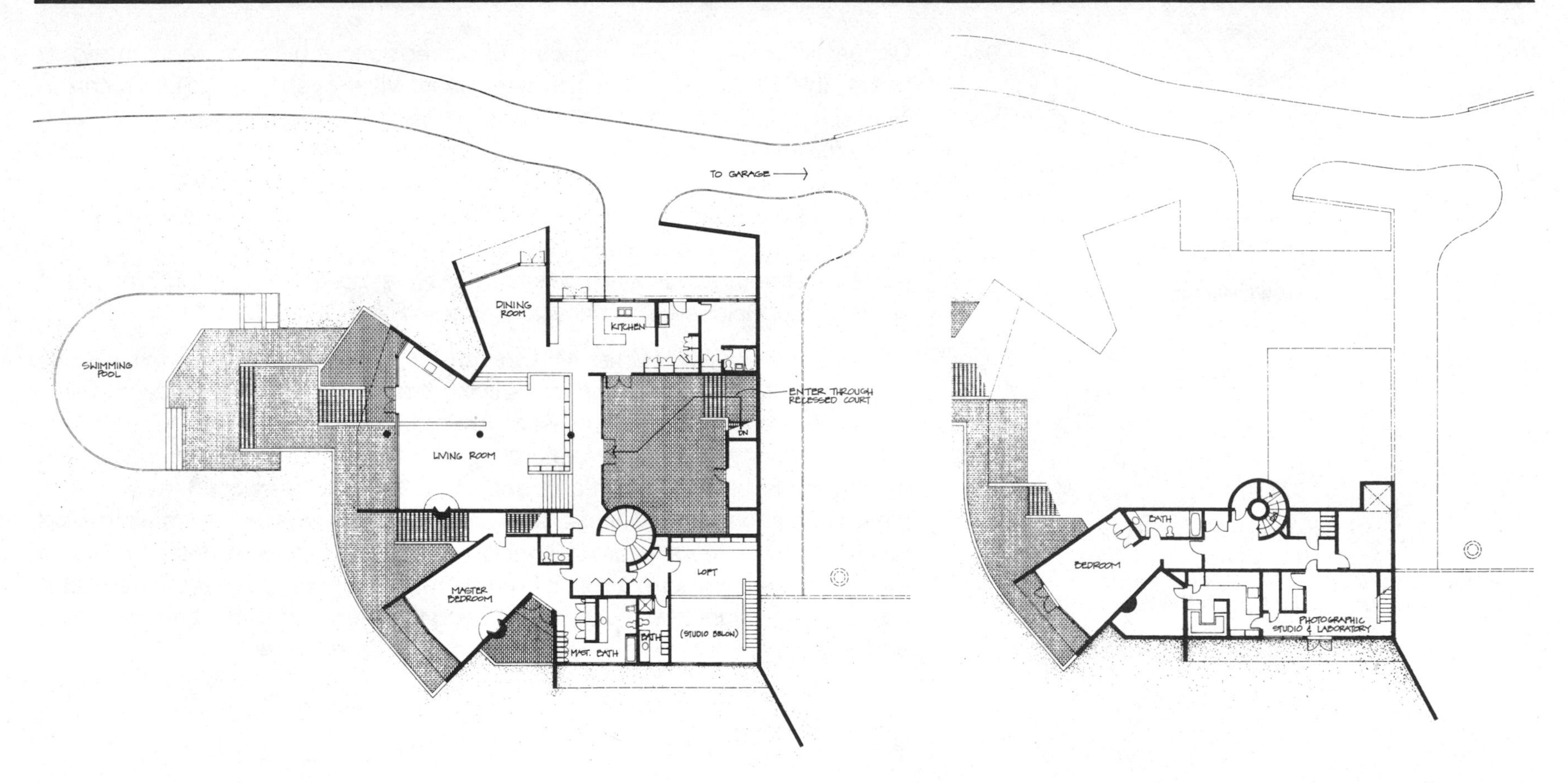

middle level floor plan

lower level floor plan

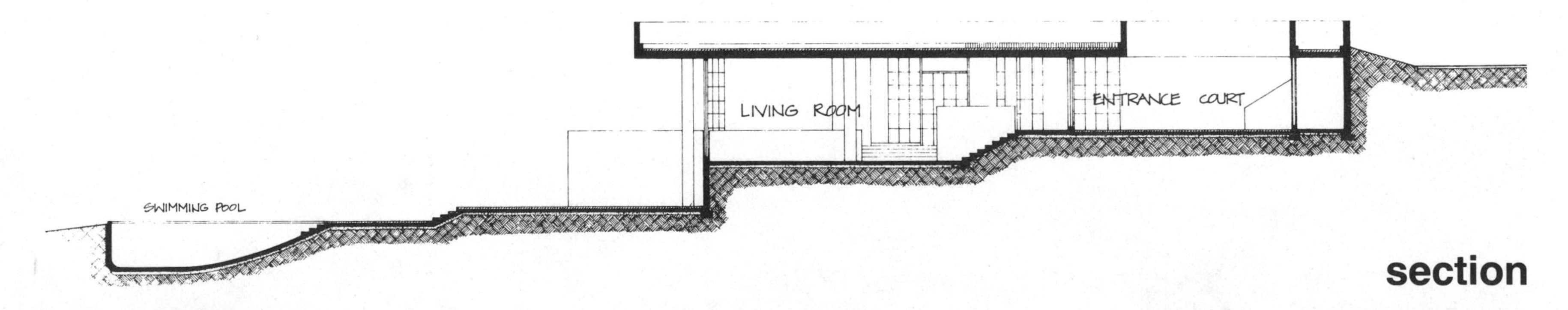

section

geier house

This house may be the first earth covered dwelling of the post-war period that was designed purely on its own merit, and not as a response to the cold war or other "external" reasons for building below the surface. It predates our current conservation interest in underground structures, and represents instead an exploitation of the earth as an element of architectural form.

Architect Philip Johnson, has created an unusual and interesting integration of earth and water into the Geier house design. All major living spaces have extraordinarily large window areas, all with controlled views over the lake and outside terraces. Because the site itself was relatively devoid of features of visual interest other than the lake, Johnson made the house and lakeshore a single sculptural unit, capped with the abstract geometric forms of weathering steel cylinders on the roof. These conceal chimneys, mechanical equipment, and roof skylights.

This house is the product of a deliberate and imaginative manipulation of the earth; it illustrates clearly that subsurface design is not some type of "non-building", but that it possesses all the qualities and opportunities for great architecture.

After twelve years of experience living in this house, the inhabitants report that "it's an absolute joy to live in."

LOCATION: Cincinnati, Ohio
ARCHITECT: Philip Johnson, 375 Park Ave., NY, NY 10022
ENGINEERS: Baxter, Hodel & Donnelly & Preston, Cincinnati, Ohio
CONSTRUCTED: 1965
PHOTOGRAPHY: Ezra Stoller, Mamaroneck, NY
Louis Checkman, Jersey City, NJ

GROSS AREA: 4,000 sq. ft.
EARTH COVER: 15 in.
STRUCTURE: 6 in. reinforced concrete roof
12 in. reinforced concrete walls
INSULATION: 2 in. foamglas, roof & walls
WATERPROOFING: Roof—5 ply built up roofing
Walls—3 ply built up waterproofing
Floor—3 ply built up waterproofing
HEATING SYSTEM: 4 area furnaces

geier house

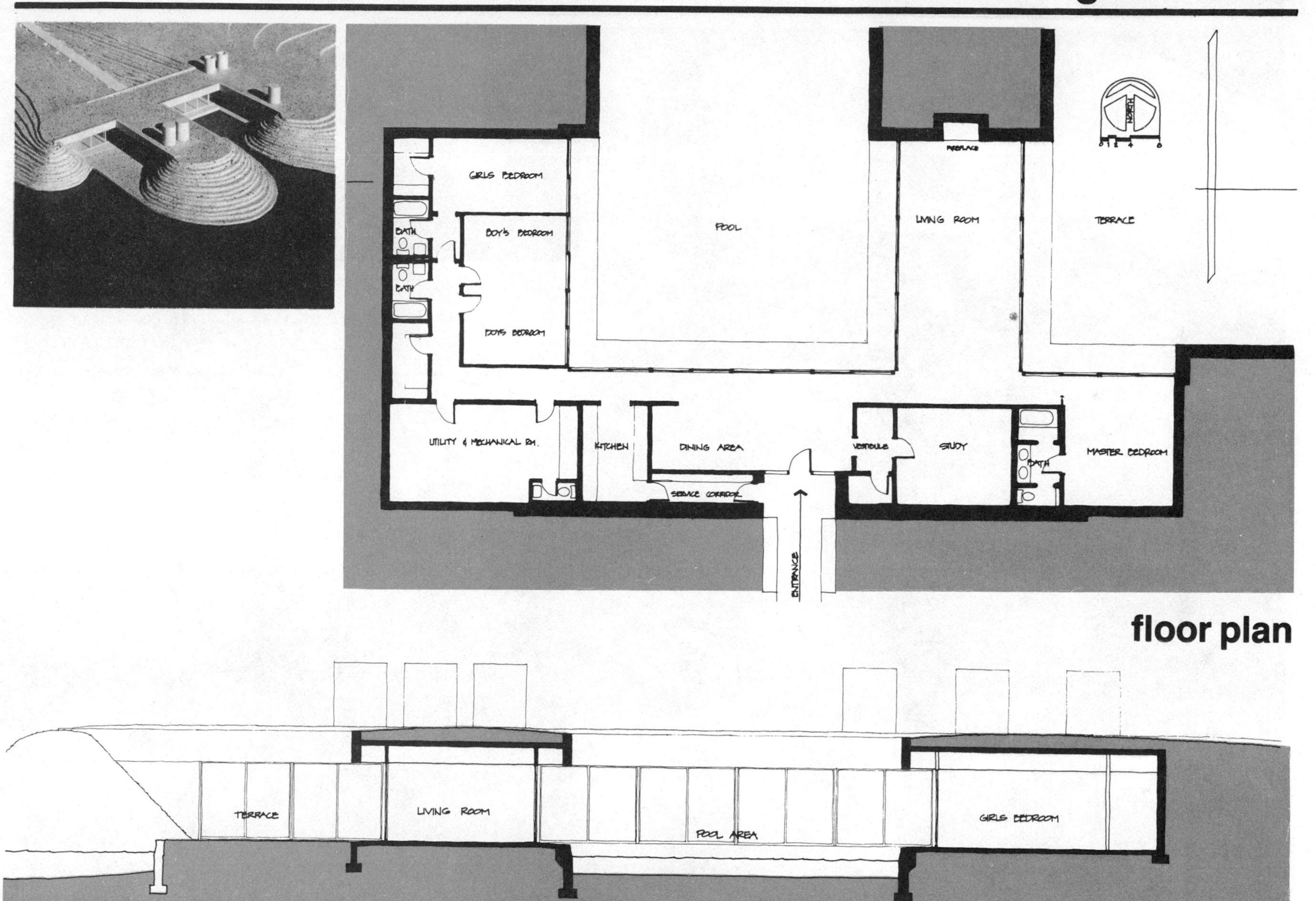

floor plan

section

geier house

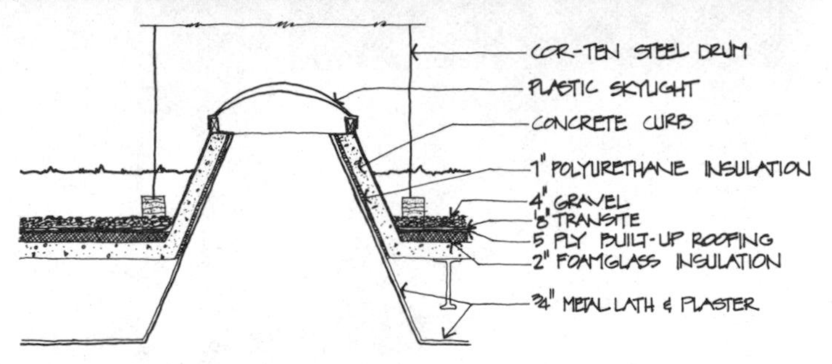

skylight detail

LOCATION:	(Not built)
ARCHITECT:	Herman DeJong Richard S. Cowan Assoc., Quakertown, Pa.
GROSS AREA:	1,600 sq. ft.
EARTH COVER:	24 in.
STRUCTURE:	8 in. concrete block walls 10 in. precast concrete plank roof concrete slab floor
INSULATION:	Roof—foamglass tapered from 4 in. to 2 in.
WATERPROOFING:	1 in. butyl membrane
HEATING SYSTEM:	Solar collectors, wood stove back up system

Terrasol #1 is one of three prototypical earth covered, solar heated houses designed by architect Herman DeJong. DeJong is a staunch supporter of energy conservation design, and of alternative energy sources. The earth protected Terrasol series of designs is a first step towards self-sufficient housing; they are designed integrally around a south facing wall of solar panels, and a berm sheltered north wall that provides internal storage areas. Wood burning stoves would be sufficient to augment the solar system in most climates, and in many areas the additional provision of a windmill-driven generator could attain nearly total energy autonomy for the house.

All three Terrasol designs prepared by DeJong are suited for flat sites, although they could be readily adapted to southern sloping hillsides. The emphasis is on practicality, and the architect feels that his houses are competitive with conventional suburban units. Construction involves simple spans of precast concrete roof planks between parallel masonry bearing walls. The kitchen and two smaller bedrooms have windows penetrating the south wall, although the major window areas of the house occupy the living room and master bedroom. These face into private outdoor terraces which are sheltered from northerly winds.

Information about the Terrasol prototypes may be obtained from Herman DeJong, AIA, Sun Wind Company of Pennsylvania, Box 16, Fairview Village, PA 19409.

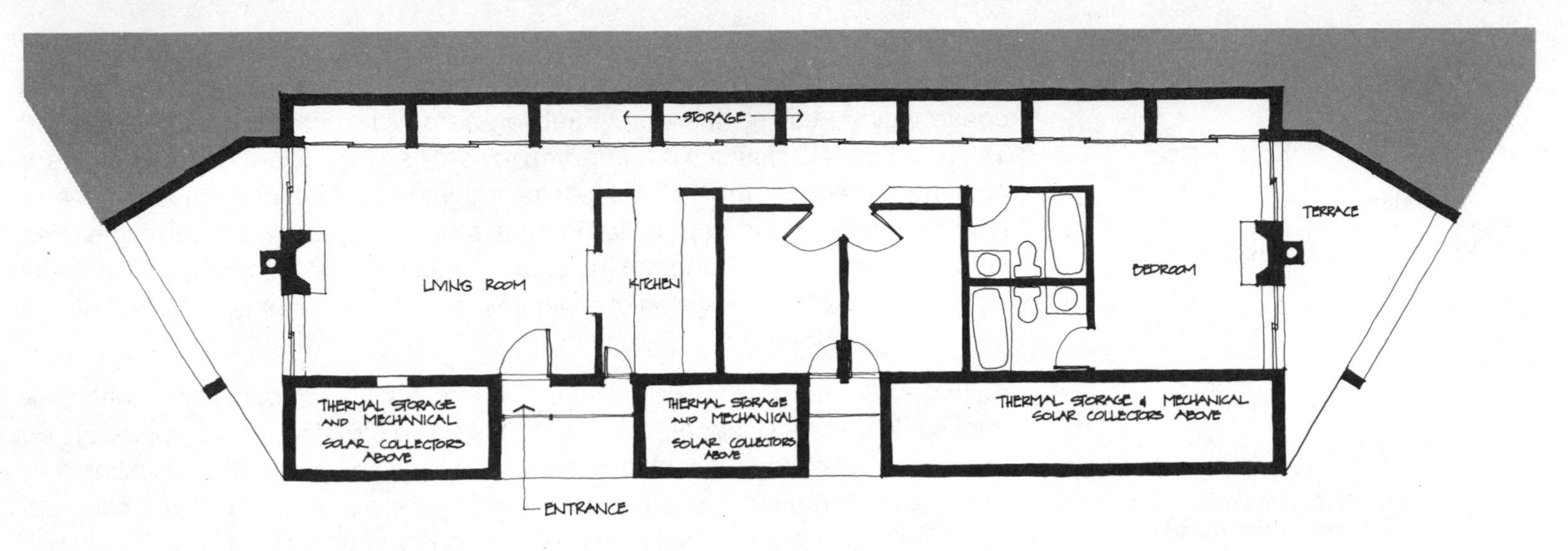

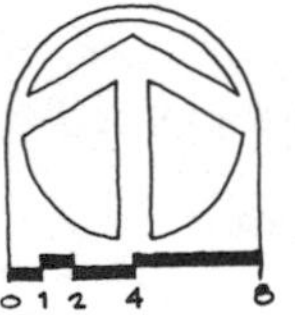

floor plan

section

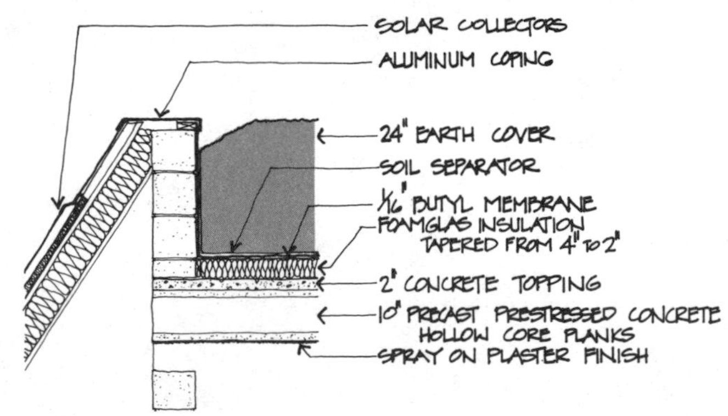

roof detail

part c:
additional information

1 appendices

appendix A energy calculations: earth sheltered building

Due to the thermal capacitance of the surrounding environment of an earth sheltered structure, conventional methods of calculating the heat flux through the walls by means of steady state approximations to prevailing weather conditions simply do not provide an adequate description of the building's thermal performance. The inherent time lag in the response of a high mass structure can serve to delay any appreciable change in the interior conditions for a day or longer after the occurrence of the event and also tends to dampen the severity of the change. Thus, seasonal extremes in air temperature are not sufficient criteria in the load analysis as the thermal behavior of the soil also reflects the history of preceding weather patterns leading up to those extremes.

To provide the time history of temperature fluctuations at the ground surface, a transient numerical analysis was carried out for earth sheltered structures by means of the Cyber 74 computer facility at the University of Minnesota.

The general equation for heat conduction, neglecting internal heat generation, is given as:

$$\rho \, C_p \frac{\delta T}{\delta t} = \overline{\nabla} \cdot (k \nabla T) \qquad \text{(A-1)}$$

where ρ is the local density of the medium (kg/cu m), Cp the specific heat (J/kgK), and k is its thermal conductivity (W/mK). T is the temperature and t represents time expressed in units consistent with the other parameters. In order to provide a means of reducing the computing time and storage requirements so that individual construction details could easily be isolated and examined closely, two programs were written for the analysis. The first was based upon the one-dimensional form of equation A-1:

$$\rho \, C_p \frac{\delta T}{\delta t} = \frac{\delta}{\delta x} (k \frac{\delta T}{\delta x}) \qquad \text{(A-2)}$$

and provided a convenient and effective means of examining the roof structure.

By means of this first program, various combinations of soil and insulation cover upon a precast concrete support structure were analyzed and compared both on an hourly basis to determine their diurnal characteristics and on a day to day basis throughout the year to provide a knowledge of their seasonal characteristics.

The second program was based on the two-dimensional heat conduction equation:

$$\rho \, C_p \frac{\delta T}{\delta t} = \frac{\delta}{\delta x} (k \frac{\delta T}{\delta x}) + \frac{\delta}{\delta y} (k \frac{\delta T}{\delta y}) \qquad \text{(A-3)}$$

with which cross sections of the entire building structure were examined on a month to month basis throughout the year.

A finite difference approximation was used to describe equations (A-2) and (A-3) with variable grid spacing to permit greater precision at the interfaces without incurring excessively large numbers of grid points in the bulk of the surrounding soil. The density and specific heat were specified for each grid point while the harmonic means for the thermal conductivity at the control volume interfaces were utilized in the right hand side of the equation. The advantage of the harmonic mean for thermal conductivity over a simple arithmetic mean is because of the nature of the heat conduction between two grid points being controlled by the series connection of two thermal resistances. In the case of variable grid spacing, however, it should be noted that it is preferable to define control volume faces as being located midway between adjacent grid points in order to avoid the cumbersome need to insert geometric factors into the averages. At this point, a second important feature is worthy of attention. By constructing the difference equations according to equations (A-2) and (A-3) and not expanding the spacial derivatives further, the task of varying soil properties between time steps is greatly simplified. This is of particular value where it is desired that the effects of moisture migration be included in the analysis.

The boundary conditions at the ground surface and on the inside building surfaces were prescribed by the convection boundary condition expressed by an energy balance on the surface control volume:

$$\rho \, C_p \, V \frac{\delta T}{\delta t} = Q_{conc} - h \, A_{surf} \, (T - T_{air}) \qquad \text{(A-4)}$$

where V is volume and Asurf is the area exposed to convection losses. The term on the left represents the change in stored energy within the control volume. The first term on the right hand side represents gains by conduction through the control volume faces and the second term prescribes convection to the adjacent air. The finite difference formulations for equation (A-4) were then modified to suit the particular geometry of each surface control volume. The convection heat transfer coefficient, h, at the ground surface is consistent with the National Weather Bureau measurements of a 15.29 km/hr (9.5 mph) average wind velocity. On the interior surfaces of the structure, a value of 2.29 W/sq m-K (0.40 BTU/hr-sq ft-°F) was used for the film conductance.

The outside air temperature throughout the year was taken from averages based on Weather Bureau data over the 30 year period from 1940 to 1970. In an effort to represent as extreme a condition as would be reasonable for the Minneapolis-St. Paul area, the summertime air temperatures were adjusted to the daytime averages. Air temperature within the building was allowed to attain the same temperature. This is primarily responsible for the lower summer cooling rates than might be expected as the internal heat generation of the building would raise this temperature to 25.6°C (78°F) during the summer months, thus resulting in a greater heat loss to the walls than is indicated by the 23.6°C (74.5°F) average outdoor air temperature.

The building structure was assumed to be constructed entirely of precast concrete beams with an effective thermal conductivity of 0.45 W/mK (0.26 BTU/hr-ft°F). This is somewhat less than the 0.81 to 1.40 W/mK (0.47 to 0.81 BTU/hr-ft°F) which would result from the use of poured concrete. However, as the concrete structure contributed only a small percentage to the total building insulation, this difference would account for less than a 4.5% change in the roof structure performance and is negligible for the walls and floor. Therefore, the general conclusions of this study are not affected by this choice.

The soil chosen for this study was a sandy loam indigenous to southeastern Minnesota. The thermal conductivity of 0.81 W/mK (0.47 BTU/hr ft-°F) is indicative of a very dry soil. The specific heat and density employed for the study were 858 J/kgK (0.205 BTU/lb-°F) and 1815 kg/cu m (113 lb/cu ft) respectively. As moisture migration in soils remains an extremely complex and little understood phenomena, it was decided that a constant moisture content would be used throughout the study to avoid the possibility of obscuring the pertinent results. In studying a single case with moisture contents ranging from 18% to 37%, however, it was found that the winter heat loss was increased by 28%, while the summer cooling rate increased 49% for the wet soil as opposed to the dry. While such variations show little impact upon the general conclusions with respect to the advantages of earth sheltered housing to conventional housing, they do emphasize the impact the moisture content has on the thermal characteristics of the soil.

appendix B energy calculations: above grade building

The energy use calculations for the above grade houses shown in the energy analysis section of the study are based on a number of assumptions. The four houses included under Case B all are assumed to be slab-on-grade construction (without a basement). Walls are of frame construction with wood exterior siding, gypsum board interior finish and fiberglass batt insulation. The ceiling/roof is assumed to be a flat, non-ventilated wood structure with pitch and gravel cover, gypsum board interior and fiberglass batt insulation. The overall dimensions of the houses in Case B are:

North, south wall length	14 meters (45.93 ft)
East, west wall length	10 meters (32.81 ft)
Ceiling height	2.5 meters (8.20 ft)

Windows are assumed to be double glazed on all four sides. The amount and location of the windows for the various houses in Case B are as follows:

Case B-1 and B-2

South side	35% of wall area	12.25 sq m (131.86 sq ft)

Case B-3 and B-4

North side	5% of wall area	1.75 sq m (18.84 sq ft)
East side	10% of wall area	2.50 sq m (26.91 sq ft)
South side	15% of wall area	5.25 sq m (56.51 sq ft)
West side	10% of wall area	2.50 sq m (26.91 sq ft)

Heat gain/loss through the windows was calculated using data from Chapter 22 of the 1972 ASHRAE Handbook of Fundamentals (referred to as ASHRAE table 2) and from table B-1 below. Note that the values for 40° and 48° north latitude were averaged to approximate values for the 44° Minneapolis latitude.

Windows are assumed to be 1.2 meters high with the bottoms at 1.0 meter above the floor. A continuous 0.9 meter overhang at the ceiling line shades the south windows. East and west windows are shaded by miscellaneous items, assumed to shade 25% of the windows. The shade factor for the windows as obtained from ASHRAE fundamentals and the window U-values are as follows:

Shade factor	0.83
U-value-daytime	3.349 W/sq m-°C (0.59 BTU h/sq ft-°F)
U-value-nighttime	2.555 W/sq m-°C (0.45 BTU h/sq ft-°F)

Nighttime U-values are lower than daytime values because drapes are assumed to be drawn from dusk to dawn.

Heat gain through the south windows was calculated using the solar intensities given in ASHRAE Table 2. The percent of the south windows in sunlight for each hour of daylight was calculated from solar altitude angles given in the table. Sunlit portions were calculated directly using the south window values in ASHRAE Table 2, while shaded portions were calculated using north window values. The formula used is as follows:

Heat Gain = sum of solar intensity x shade factor x days per month.

	average monthly outdoor temp.	average monthly night time outdoor temp.	average monthly day time outdoor temp.	percent of possible sunshine
january	−11.0°C	−12.2°C	− 8.5°C	51%
february	− 8.6	− 9.4	− 6.7	51
march	− 2.1	− 3.8	− 0.4	54
april	7.3	5.0	9.6	55
may	13.9	10.3	16.6	58
june	19.4	15.8	21.9	63
july	22.2	18.1	25.2	70
august	21.2	18.6	23.8	67
september	15.6	13.4	18.7	61
october	10.0	8.7	11.8	57
november	0.2	0.3	1.7	39
december	− 7.4	− 8.2	− 5.9	40

b-1 average outdoor temperatures and percent of possible sunshine.

Heat gain for east and west windows was calculated in a manner similar to that used for south windows. Heat gain for north windows was calculated in a manner similar to that used for south windows except no portion of the north window area is assumed to be in direct sunlight. Heat gains for all windows during cloudy periods were assumed to be 10% of the sunny period values. Heat losses through all windows were calculated with the U-values shown above and the outdoor temperatures shown in Table B-1.

Heat gain/loss through walls for months except June, July and August was calculated from the formula $Q_R = Q_H + Q_C$, which says that the radiant heat gain to the surface of the wall equals the net total of the heat convected away from the surface of the wall to the air plus the heat conducted from the surface of the wall to the interior of the building. The value for H was derived from information in the 1972 ASHRAE fundamentals and is as shown below. U-values for the conduction from the exterior wall surface to the interior of the building are the same as the wall U-values less the exterior air film coefficient. Calculations were made for two different walls called X and Y. Wall X is used in Cases B-1 and B-3 while wall Y is used in Cases B-2 and B-4. Both are frame construction with 3/4 x 10 in. beveled, lapped siding, 25/32 in. sheathing and 1/2 in. gypsum board on the interior. The X wall is constructed of 2 in. x 4 in. lumber filled with 3 1/2 in. fiberglass batt insulation. The Y wall is constructed of 2 in. x 6 in. lumber filled with 5 1/2 in. fiberglass batt insulation.

E = 0.204 sq m/hr (0.019 sq ft/hr)
H = 29.878 W/sq m-°C (5.263 BTU h/sq ft-°F)
UX = 0.386 W/sq m-°C (0.068 BTU h/sq ft-°F)
UY = 0.255 W/sq m-°C (0.045 BTU h/sq ft-°F)

Heat gains for June, July and August were calculated assuming that all of the radiant heat absorbed by the exterior surface of the wall was conducted into the building. The emissivity (E) of the wall is as shown above. Heat losses for June, July and August were disregarded.

Radiant heat gains to the surface of the walls were corrected for shading by the south overhang and east and west miscellaneous items and for cloudy periods in a manner similar to that for windows. Radiant heat gains during cloudy periods were assumed to be zero. Heat gain/loss through roofs was calculated in a manner similar to that used for the walls with an X roof and a Y roof. Roof X is used in Cases B-1 and B-3 while roof Y is used in Cases B-2 and B-4. No shading of the roof was assumed. Both are frame construction with pitch and gravel topping, 2 in. wood deck and 1/2 in. gypsum board on the interior. The X roof is constructed of 2 in. x 6 in. lumber with 5 1/2 in. fiberglass batt insulation. The Y roof is constructed of 2 in. x 10 in. lumber with 9 in. fiberglass insulation. Both roofs are flat and unventilated. The following values were used in the roof calculations.

E = 0.258 sq m/hr (0.024 sq ft/hr)
H = 29.878 W/sq m-°C (5.263 BTU h/sq ft-°F)
UX = 0.261 W/sq m-°C (0.046 BTU h/sq ft-°F)
UY = 0.182 W/sq m-°C (0.032 BTU h/sq ft-°F)

Heat loss through the floor slab was calculated using the ASHRAE fundamentals method. A value of 43 BTU h per foot of exposed slab edge was used for the maximum loss. Monthly values were assumed to be in the same ratio to this value as the ratio of the temperature differentials. Summer values were disregarded.

For Case F, the above grade house was a basement, all of the same basic data applies which has been presented for Case B. Case F-1 uses wall and roof X while Case F-2 uses wall and roof Y. In both cases, there is 35% glass area on the south wall with no windows on the north, east and west walls. The only difference between Case B and Case F is that the floor slab is replaced by an earth sheltered basement. The values for the basement area are taken from the calculations for earth sheltered walls and floors.

In all of the earth sheltered cases (A, C, and E) the exposed elevation is assumed to be wall Y and construction with 35% window area.

appendix C energy comparisons: earth sheltered vs above grade

The material presented in this appendix is intended to provide a more detailed breakdown of the energy use comparisons presented in the energy analysis section of the report. The figures presented here are for the sub-grade chamber (Case D) and the surface slab-on-grade house (Case B). These are shown since they are the simplest cases and the other cases are based on these calculations. It is important to note that the data presented here is based on a number of assumptions and criteria presented in Appendices A and B. These limitations must be recognized before applying this data to other situations.

The construction of the surface building components is as described in Appendix B. The sub-grade chamber construction is as follows:

roof: 30 cm precast concrete
10 cm extruded polystyrene insulation
50 cm soil

walls: 30 cm concrete
10 cm extruded polystyrene insulation covering top half of wall

floor: 30 cm concrete

Table C-1 indicates the heat loss/gain through the windows of the above grade structure.

Table C-2 compares the energy requirements of the walls of the subgrade chamber and the surface structure. Values for both standard (type X) and well insulated (type Y) walls are given. Different values for north, east, west, and south walls must be given for the surface structure due to the effect of solar radiation on the various walls.

Table C-3 compares the heat loss/gain through the roofs of the subgrade chamber and the surface structure (with type X and type Y construction).

Table C-4 indicates the values used for calculation of the heat loss through the floors of the two structures. No direct comparison can be made because of the differences in the methods of calculation, however they are presented so that all of the major building components used in the overall energy use totals are included.

In all of the tables, totals are given for the summer and the winter. The grand total for the entire year is a meaningless value for an above grade low-mass structure. There is no carry-over from summer to winter or winter to summer. A better indication of yearly energy requirements is a total of the absolute summer and winter values. As calculated values are shown in the tables. These values have been rounded before incorporation into the body of the report.

month	north	east/west	south
january	−60.37	−48.95	+ 3.90
february	−48.57	−31.48	+17.53
march	−36.84	−11.02	+14.78
april	−12.09	+16.34	+ 9.39
may	+ 9.11	+37.43	+11.80
june	+26.66	+55.77	+23.45
july	+25.87	+60.11	+27.90
august	+18.26	+51.12	+41.62
september	+ 3.19	+28.04	+55.74
october	−13.59	+ 3.99	+53.82
november	−35.42	−26.68	+13.53
december	−54.61	−47.08	− 5.56
june thru aug. totals	+70.79	+167.00	+92.96
oct. thru april totals	−261.94	−144.98	+107.40

note: positive values indicate heat flow into building, negative values out.

c-1 heat gain/loss through windows—kW-hr/sq m

		one level sub-grade cham. (4 walls)	X-wall			Y-walls		
			north	east/west	south	north	east/west	south
january		−1.91	−8.90	−8.90	−8.89	−5.89	−5.89	−5.88
february		−2.32	−7.42	−7.42	−7.41	−4.91	−4.91	−4.90
march		−2.57	−6.33	−6.32	−6.32	−4.19	−4.19	−4.18
april		−2.61	−3.53	−3.53	−3.53	−2.34	−2.34	−2.33
may		−2.45	−1.73	−1.73	−1.73	−1.15	−1.15	−1.14
june		−4.71	+0.45	+0.95	+0.46	+0.30	+0.63	+0.30
july		−3.68	+0.45	+1.04	+0.54	+0.30	+0.69	+0.36
august		−3.09	+0.31	+0.88	+0.65	+0.21	+0.58	+0.42
september		−0.17	−1.24	−1.23	−1.22	−0.82	−0.81	−0.81
october		−0.69	−2.87	−2.87	−2.86	−1.90	−1.90	−1.89
november		−1.01	−5.50	−5.49	−5.49	−3.64	−3.64	−3.63
december		−1.42	−7.88	−7.88	−7.88	−5.22	−5.22	−5.22
june thru aug.	total	−11.48	+1.22	+2.87	+1.65	+0.81	+1.89	+1.09
oct. thru april	total	−12.54	−42.44	−42.41	+42.38	−28.09	−28.07	−28.05

note: positive values indicate heat flow into building, negative values out.

c-2 heat gain/loss through walls—kW-hr/sq m: chamber, standard insulation (X), well insulated (Y)

		sub-grade chamber U=0.217 W/sq m−C°	"X" roof-surface U=0.261 W/sq m−C°	"Y" roof-surface U=0.182 W/sq m−C°
january		−4.30	−6.02	−4.19
february		−4.16	−5.01	−3.49
march		−3.39	−4.27	−2.97
april		−2.17	−2.37	−1.65
may		−1.82	−1.15	−0.80
june		−1.44	+3.20	+2.22
july		−0.52	+3.48	+2.42
august		−0.53	+2.86	+1.98
september		+0.08	−0.82	−0.57
october		−1.18	−1.93	−1.34
november		−2.53	−3.71	−2.58
december		−3.68	−5.33	−3.71
june thru aug.	total	−2.50	+9.54	+6.62
oct. thru april	total	−21.40	−28.63	−19.92

note: positive values indicate heat flow into building, negative values out.

c-3 heat gain/loss through roof kW-hr/sq m

		sub-grade (kW-hr/sq m)	slab X & Y (kW-hr/m of edge)
january		−0.71	−67.49
february		−0.74	−56.26
march		−0.78	−48.02
april		−0.81	−26.80
may		−0.83	−13.18
june		−2.84	—
july		−2.28	—
august		−2.14	—
september		−0.04	− 9.36
october		−0.44	−21.77
november		−0.58	−41.67
december		−0.64	−59.75
june thru aug.	total	−7.26	—
oct. thru april	total	−4.70	−321.76

note: positive values indicate heat flow into building, negative values out.

c-4 heat gain/loss through floors

Only the code provisions which entail special considerations for earth sheltered houses will be discussed here. The code referred to is the Uniform Building Code 1976 Edition and 1977 Supplement. Where the Minnesota State Building Code (S.B.C.) has substantially modified these provisions, this will be noted.

part I administrative

UBC Section 106 Alternate Materials and Methods of Construction
SBC Section 103

This section allows the use of materials and methods of construction that have not specifically been prescribed by the code. The design for these materials and methods must comply with the general design requirements in the code and the materials or method must be the equivalent of that prescribed in the Code in quality, strength, effectiveness, fire resistance, durability and safety.

The onus of proof for the acceptability of the material or method is with the prospective builder.

UBC Section 240 Board of Appeals

This section establishes a Board of Appeals to consist of five members. The Board is given the leeway to adopt reasonable rules and regulations for conducting its investigations and may recommend new legislation to the governing body. The Board will give all decisions and findings in writing to the Building Official with a duplicate copy to the appellant.

This would be the body which could grant variances from some of the specific building code requirements such as egress from sleeping areas. Questions could be referred to the variance board if the local code official felt that an alternate method of construction was outside his responsibility to decide on or if the code official disallowed a request and the prospective builder made an appeal.

part II definitions and abbreviations

chapter 4 definitions and abbreviations

Some of the pertinent definitions from the Uniform Building Code are given below.

Basement—is that portion of a building between floor and ceiling which is partly below and partly above grade (as defined in this Chapter), but so located that the vertical distance from grade to the floor is less than the vertical distance from grade to ceiling.

Cellar—is that portion of a building between floor and ceiling which is wholly or partly below grade (as defined in this Chapter) and so located that the vertical distance from grade to the floor below is equal to or greater than the vertical distance from grade to ceiling.

Grade (Adjacent Ground Elevation)—is the lowest point of elevation of the finished surface of the ground, paving or sidewalk within the area between the building and the property line, or wnen the property line is more than 5 feet from the building, between the building and a line 5 feet from the building.

Habitable Space (Room)—is space in a structure for living, sleeping, eating or working. Bathrooms, toilet compartments, closets, halls, storage or utility space, and similar areas are not considered habitable space.

Smoke Detector—is an approved detector which senses visible or invisible particles of combustion. The detector shall bear a label or other identification issued by an approved testing agency having a service for inspection of materials and workmanship at the factory during fabrication and assembly.

Story—is that portion of a building included between the upper surface of any floor and the upper surface of the floor next above, except that the topmost story, shall be that portion of a building included between the upper surface of the topmost floor and the ceiling or roof above. If the finished floor level directly above a basement, cellar or unused under-floor space is more than 6 feet above grade as defined herein for more than 50 percent of the total perimeter or is more than 12 feet above grade as defined herein at any point, such basement, cellar or unused underfloor space shall be considered as a story.

NOTE: The term grade in the definition of a basement or cellar is intended to refer to the average grade (as defined here) around the building perimeter.

part III requirements based on occupancy

The Uniform Building Code has three major sets of requirements:

1. Requirements based on occupancy.
2. Requirements based on location in fire zones.
3. Requirements based on types of construction.

These three sets cover the major differences in life hazard from one building to another. The requirements based on occupancy differentiate, for instance, between buildings used for storing hazardous materials, buildings where large groups of people meet and single family residences. These clearly have very different life safety requirements. The requirements based on location in fire zones address mainly the prevention of fire spread from one building to another in a congested area. The requirements based on types of construction are intended to insure that only suitable materials and methods of construction are used for the various occupancies and fire zones. These requirements are hence intertwined very closely with the first two sets.

A building is classified according to its occupancy by placing it in one of the following building groups and divisions:

Group A Assembly buildings with varying occupancy loads and intrinsic fire hazards.
Divisions 1,2, 2.1, 3 and 4
Group E Educational buildings with varying degrees of size and use.
Divisions 1,2,3
Group I Hospitals, Nursing Homes and Nurseries.
Divisions 1,2,3
Group H Storage and handling of hazardous materials.
Group B Industrial and commercial buildings with no special hazards.
Divisions 1,2,3,4
Group R Residential occupancies ranging from hotels to single family dwellings.
Divisions 1,3
Group M Miscellaneous occupancies such as private garages, sheds and agricultural buildings.

For full definitions of each of the occupancies, reference should be made to the building code. Single family dwellings which are the main concern of this manual would all be classed under Group R Division 3 (R-3) Occupancies which are defined as dwellings and lodging houses.

chapter 14 requirements for group R division 3 occupancies (housing).

Section 1402 Construction, Height and Allowable Area

R-3 Occupancies are essentially unlimited in area, but are limited in height to three stories except for buildings built in Type 1 Fire Resistive Construction (see Part V), which are unlimited in height also.

Sections 1403 Location on property
1406 Yards & Courts

The requirements on fire resistance of exterior walls and openings in exterior walls applicable to R-3 occupancies are shown in the table below. Almost all dwellings will be in Fire Zone 3 and hence, when the exterior wall is more than 3 feet from the property line, no requirements apply. When interior courts have windows required for light or egress opening onto them, the following dimensions shall apply. The court shall not be less than 3 feet in width. With windows on opposite sides of a court, the minimum width shall be 6 feet. Courts bounded on three or more sides by the walls of the building shall not be less than 10 feet in length unless bounded on one end by a street or yard. Yards and courts more than two stories in height have some additional regulations which will not be considered here. Eaves and cornices should not project into any required yard more than 2 inches for each foot of yard width. Other appurtenances such as unroofed landings and stairs may project into the required yard provided no portion extends above the floor level of a habitable room.

fire zone	fire resistance of exterior walls	openings in exterior walls
1	2 hr. less than 20 ft. 1 hr. elsewhere	not permitted less than 3 ft. protected less than 20 ft.
2	1hr.	not permitted less than 3 ft. protected less than 20 ft.
3	1 hr. less than 3 ft.	not permitted less than 3 ft.

Section 1404 Exit facilities

For the house in general, stairs and exits must comply with Chapter 33 of the U.B.C. Only one exit is required for dwellings with an occupant load of less than 10 people. 300 square feet per occupant is used in determining the occupant load, and hence, a house of less than 3,000 square feet only requires one exit. The exit referred to here is a means of egress to a public way including intervening doors, corridors, courts, and yards, etc. There are many provisions for maximum distances to exits, stairways requirements, etc. which apply to all buildings. When considering earth sheltered dwellings in particular, few of these requirements have any significance. The major pertinent requirements are:

1. The maximum distance of travel from any point to an exterior exit door shall not exceed 150 feet.
2. Exits may pass through an intervening room or area provided this gives a direct means of egress to the outside or to exit stairways or corridors as defined in the code.
3. Exits must not pass through kitchens, storerooms, restrooms, closets, etc.
4. Exit doors shall not be less than 3 feet in width and 6 feet 8 inches in height and shall open at least 90 degrees. The clear width when open shall be not less than 32 inches.
5. Doors must be openable from the inside without the use of a key.
6. Where the occupant load is over 10 persons, exit corridors must be at least 44 inches wide and 7 feet high.
7. Stairways for an occupant load of less than 10 must be at least 30 inches wide.
8. In exit stairways for dwellings with an occupant load less than 10, the rise shall not

exceed 8 inches and the run shall not be less than 9 inches.

9. Winding and spiral stairways are allowed to be used as exit stairways in dwellings provided they meet certain requirements.

For sleeping rooms, extra exit requirements apply and the code requires at least one operable window or exterior door in each sleeping room. Such windows must have a net clear opening of 5.7 square feet with a minimum net height of 24 inches and a minimum net width of 20 inches. The finished sill height of the window must be less than 44 inches above the floor.

Section 1405 (a) Light and Ventilation

All habitable rooms (see definition), within a dwelling unit must be provided with natural light by means of exterior glazed openings. The minimum area is one-tenth of the floor area of the room or 10 square feet whichever is less. The exterior glazed openings may include all exterior windows or skylights in a room. The requirements are based on the nominal dimensions of the window or skylight and not on its net glass area. There are provisions for natural ventilation requiring that openable exterior openings of a least one-twentieth of the floor area of habitable rooms be provided. Bathrooms, laundry rooms etc. have a further minimum requirement of exterior openings of 1 1/2 square feet and habitable rooms have a minimum of 5 square feet. For the ventilation requirements only, the requirements can be waived if a mechanical ventilating system is provided. The mechanical ventilation requirements are shown in the table below.

room	air movement	outside air
habitable rooms	2 air changes/hr.	1/5 of supply from outside
bath, laundry rooms, etc.	5 air changes/hr. capability	direct to outside

When considering light and ventilation requirements, rooms can be combined when at least 50% of the area of the common wall is open and unobstructed. There is a further requirement to prevent a very small common wall from being used. This requires that the opening be greater than one-tenth of the floor area of the interior room with a minimum size of 25 feet.

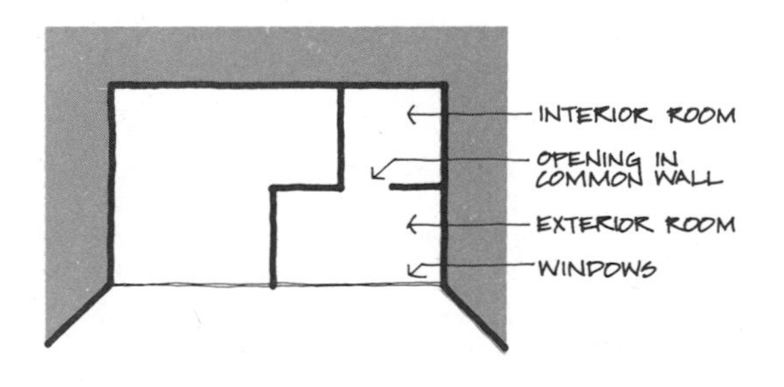

The code further states that required openings for natural light and ventilation open directly onto a street or public alley or a yard or court located on the same lot as the building. This is to prevent a building later being built on an adjacent lot and cutting off natural light and ventilation. The minimum yard and court requirements for **required** openings were given in the discussion of sections 1403 and 1406. Required windows are permitted to open onto a roofed porch provided its ceiling height is not less than 7 feet and its longer side is at least 65 percent open.

Section 1407 Room dimensions

These dimensions apply to any dwelling and have no special significance for earth sheltered dwellings. The major requirements are listed below.

Ceiling heights— Habitable rooms 7 feet 6 inches.
Height required in two-thirds of the room area.
No flat ceiling less than 7 feet.

Sloped ceilings—Above height required in one-half of the room area.
Area of room with height less than 5 feet shall not be included in any minimum area computation.

Floor area— At least one room greater than 150 square feet.
Minimum area of rooms 70 square feet (except kitchen).

Width— Minimum room width 7 feet (except kitchen).

Section 1410 Heating

The heating system must be capable of maintaining a room temperature of 70° F at 3 feet above the floor in all habitable rooms.

part iv requirements based on location of fire zones

Fire Zones are established by a city to identify the areas of the city where severe hazards of fire spread from one building to another exist. Three fire zones are labeled by the Code as Fire Zones, 1, 2 and 3. It is up to the city to deine the extent of each zone. A Fire Zone No. 1 would be, for instance, the downtown area of a large city, a Fire Zone No. 3 would be residential and outlying areas and Fire Zone No. 2 an intermediate classification. Since these guidelines are directed at dwellings which are unlikely to be located in Zones 1 and 2 and since Fire Zone No. 3 has no specific restrictions on construction type, this section will not be reviewed in any greater detail.

part v requirements based on types of construction

The types of construction defined in the building code are given below.

Construction type	Example	Comment
Type I F.R.	Concrete frame building protected steel frame	Have non-combustible structure with decreasing resistance to fire in terms of length of exposure
Type II F.R.		
Type II 1 HR.	Prefabricated metal industrial building	
Type II N		
Type III 1 HR		
Type III N	noncombustible walls, light wood frame roof.	Combustible structures allowed.
Type IV H.T.	Old warehouse	
Type V 1 HR.		
Type V N	Dwelling	

Notes:

F.R. Fire Resistive

1-HR. Refers to the length of time that an element will continue to function in a fire of standard severity, i.e. a structural element should not collapse and a partition should retard the passage of fire within this period of time.

N Indicates no special fire resistive requirements beyond the minimum type requirements.

H.T. Heavy Timber Construction—this class was identified specially to cover the old form of warehouse construction. The heavy sections will only burn slowly and will retain their strength for a reasonable period of time.

These construction types were outlined for general information on how the code is set up, but when the tables outlining what types of construction are allowed for a particular fire zone or occupancy are examined, there are no meaningful requirements for dwellings in a Fire Zone 3 except the height restriction to 3 stories discussed under section 1402.

part vi engineering regulations

chapter 23 general design requirements

This chapter outlines the loadings and general design considerations to be taken into account.

chapters 24-28 masonry, wood, concrete, steel, aluminum

These chapters cover the requirements for the use of these major structural materials.

This section is where the potential home owner will need the services of a professional architect or engineer to provide the documentation that the chosen design is strong and durable enough. For conventional structures these requirements have become available in tabular form and what is an adequate construction is widely known. For earth sheltered construction, the loads are much heavier and at least initially, individually prepared designs will be almost a necessity (see structural section).

part vii detailed regulations

chapter 29 excavations, foundations and retaining walls

The provisions here which will affect preliminary design are:

1. Footings shall be of masonry or concrete unless otherwise specifically provided (wood foundations are now also approved by the I.C.B.O.) and must extend below the frost line. The Minnesota State Building Code identifies two zones with required foundation depths of 5 feet and 3 1/2 feet respectively. Where sand or gravel extends below the frost depth, construction on grade is permitted by the State Code.
2. Any permanent slopes made during construction by cutting existing ground or building up fill shall not be steeper than 2 horizontal to 1 vertical unless substantiating data including a soil investigation report are submitted.
3. When making excavations 12 feet deep or less near adjoining property, the excavation must be protected against cave-in or settlement and the owner of the adjacent property must be notified at least 10 days before the excavation. The owner of the adjacent property is, however, responsible for any protective measures necessary for his building foundations.

   When the excavation exceeds 12 feet the owner of the adjacent property is only responsible for extending his foundation to 12 feet below grade. The person causing the excavation is liable for work below this depth.

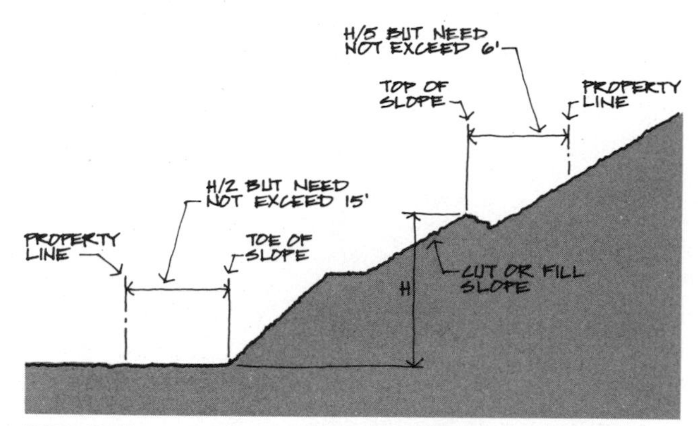

REQUIRED SETBACKS FOR PROPERTY LINES

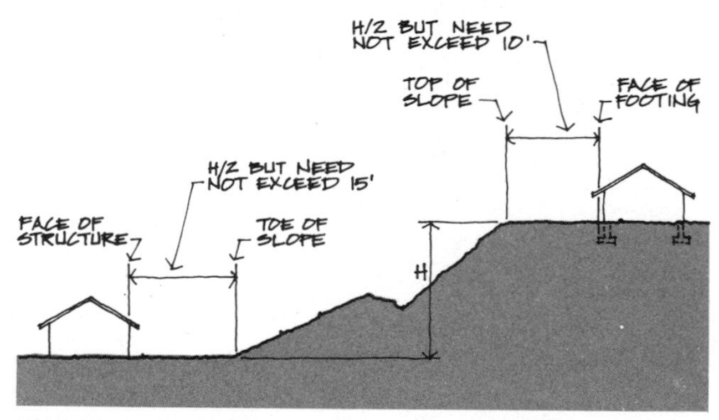

REQUIRED SETBACKS FOR STRUCTURES

appendix

chapter 70 excavation and grading

This chapter has **not** been adopted by the Minnesota State Building Code although individual cities or townships have been free to adopt it themselves. Although not law in most areas, some of the requirements will be listed to indicate for what types of grading the Code requires special attention.

This chapter requires permits for excavation and grading (with certain exceptions) and sets forth requirements for such excavation and grading. The major exceptions where a grading permit is **not** required are:

1. Where the grading is remote from private or public property.
2. Basement excavation authorized under a normal building permit.
3. An excavation less than 2 feet deep or less than 5 feet deep with a slope less than 1 1/2 horizontal to 1 vertical.
4. A fill less than 1 foot deep when supporting structures or less than 3 feet deep otherwise. The slope must be flatter than 5 horizontal to 1 vertical and the total fill must not exceed 50 cubic yards on one lot or obstruct a drainage course.

The major requirements for set-backs of slopes from property lines and buildings are illustrated in the two adjacent figures. Retaining walls may be used to reduce the required set-backs when approved. The set-backs may also be increased where soil conditions warrant it.

appendix E h.u.d. minimum property standards

101-1

New materials and technologies are encouraged but alternatives must demonstrate equivalent quality to the present standards.

101-2

Where geographic or site conditions make compliance impractical or impossible, variations to the standards may be permitted.

101-4

Variations in standards on a non-repetitive basis can be made by the field office director but for a repetitive basis substantiating data must be submitted by the field office through the regional office to the H.U.D. Office of Underwriting Standards in Washington.

205-1

Each property must have vehicular or pedestrian access from a public or private street. Private streets must have permanent easements. Seasonal homes have some relaxations on these requirements.

301-2

The site plan must be designed to fit the natural contours of the land as closely as possible and practical.

301-3

Existing good quality vegetation shall be saved wherever possible. Construction shall take place in areas on the site with the least potential ground water hazard.

303-2

The building location and arrangement shall relate well:

(a) to the natural topography, avoiding deep cuts, fills, excessive foundation wall depth, unnecessary steps and steep access gradients;

(b) to climatic conditions, assuming maximum benefit from and protection against, sun, wind, temperature, precipitation, etc.;

(c) to attractive views;

(d) to minimize undesirable noise levels.

304-2

Minimum distances from building walls to lot lines are:

(a) Primary wall $D = 6 + 2S + L/10$

(b) Secondary wall $D = 2 + S + L/10$ (Min. dist. 5 ft)

(c) Windowless wall—Dwelling D = S + L/10 (Min. Dist. 5 ft)

Detached car port or Garage
D = S + L/10 (Min. Dist. 3 ft)

(d) Retaining wall—where the top of the retaining wall is above the mid-height of adjacent habitable room windows the minimum distance is the sum of the distances for a windowless wall and the wall containing the windows.

NOTES

1. A primary wall is defined as containing the principle windows in a habitable room except bedrooms and kitchen. It also applies to the main entrance of the dwelling when it faces the primary wall of another dwelling.
2. D = minimum distance in feet
   S = the height of the wall in stories
   L = the horizontal length of the wall

Location of a windowless wall on a side lot line is acceptable to H.U.D. providing that permanent access for outside maintenance is legally secured and that the minimum distance to an adjacent dwelling is not less than the sum of the appropriate distances above (minimum 10 ft). With this exception, the minimum distance from side and rear lot lines for building walls is 5 ft.

For single family units, walls with principal windows of major daytime use rooms shall not be face to face unless privacy arrangements are made. The standards contain other qualifications of set-back distances. In particular, where the minimum slopes away from foundations terminate at the rate of a 2 on 1 or steeper slope, yard area dimensions shall be increased 1 ft. for each 4 ft. of slope height above 15 ft.

304-5

Building with grade level or low windows shall be arranged or screened to minimize headlight glare and automobile exhaust penetration into habitable rooms.

304-6

Outer courts bounded by building walls on 3 sides shall have a minimum width equal to the sum of the appropriate yard distances. This can be reduced 25 percent where the depth of the court is less than half of the width. The depth of the court shall not be greater than 1 1/2 times the width.

Inner courts bounded on all four sides by building walls are required to have an unobstructed passage of minimum size 4 ft. wide by 7 ft. high to the main yard. Sums of minimum yard distances again apply and the minimum areas allowed are 100 sq. ft. for one story of 150 sq. ft. for two stories or more.

307-2

Where the window sill of a habitable room requiring privacy is less than 6 feet above a public walk, the walk must be at least 8 feet away.

310

This section on grading design has numerous requirements for site grading including a table of maximum and minimum slope gradients. Some notes on these requirements are provided below:

1. Drain surface water away from buildings and off-site.
2. Avoid concentrating run-off onto neighboring properties.
3. Minimum slopes away from foundations
   5 percent for pervious surfaces
   1 percent impervious surfaces
4. Minimum protective slope around building is 6 inches drop and 10 feet length (or to property line).
5. First floor grades should be chosen to: avoid deep cuts and fills, allow gravity sewer service unless otherwise approved, avoid long flights of steps and minimize banks, retaining walls etc.
6. Retaining walls 24 inches or over shall have fences for visitor protection.

311

Habitable space shall be above a flood level equivalent to a 100 year return frequency.

403

Natural light requirements—minimum glazed area of 10 percent of floor area for living rooms, dining rooms and bedrooms (may be reduced to 8 percent in the near future); 1% of floor area for basements.

Ventilation—minimum openable area is 5 percent of floor area for habitable rooms, 1 percent for basements. No optional mechanical ventilation provisions given except for kitchens, bathrooms or laundry.

405

Exterior window requirements for sleeping rooms are the same as for the building code except that 2 doors providing separate ways of escape are permitted in lieu of an exterior door or window.

507

Concrete slabs on ground require a vapor barrier with a maximum vapor permeance of 1/2 perm.

607

Where habitable spaces are below grade, a waterproof membrane at least equivalent to two layers of 15 pound asphalt-saturated felt mopped between and mopped over shall be used.

Foundation walls for habitable rooms below grade must be waterproofed with at least the equivalent of 2-ply lot mopped felts, 6 mil polyvinyl chloride or 55 pound roll roofing.

Equity Requirements of Earth Covered Buildings and Instruments of Remedy*©

DEAN A. MANSON

Professor of Real Estate Finance
Southern Methodist University

When Frank Moreland invited me to participate in this conference, I thought how interesting it would be to involve some of our graduate students in a study of this nature. So we obtained from John Barnard in Massachusetts plans of his ecology house, paired off our graduate students and sent them out to savings and loan associations to determine how you would finance such a building. They all came back with the same answer, you can't. You can't, in the traditional way. So I called John Barnard in his office in Massachusetts to ask him where he financed the home he built. I called the financial institution he named and they affirmed the financing of Barnard's home. It consists of a standard 30 year mortgage, 7 1/2% interest, no points on the front end, and the payments were current. I asked if they would finance another one and they responded that if John Barnard brought them another one, they would finance it. I thought that was interesting so I called John back and talked to him again. He was very knowledgeable not only about the underground home, but also about the financial workings of the institution from which he obtained his loan. As a matter of fact, it turned out that he was on the Board of that bank. That was the closest we came to finding a bank that would finance an underground dwelling.

So we continued our investigation. We began to look through the materials available on underground homes, particularly concerning the financing. We found an absence of that kind of thing. Looking under the categories of U for underground and F for financing, we found nothing. Finally an associate of ours in Washington at the Library of Congress informed us that most of the information in the field is under C, for cave. Even John Barnard's ecology house, the plans for that being in his booklet entitled "Wentworth Institute Alumni News," which is an unusual title for an ecology house booklet, is listed under "Caves" in the Library of Congress. So apparently whenever classifiers see anything that is going underground, they put it in sort of Cro-Magnon, Neanderthal style category.

Let's presume that an underground dwelling seems to be a reasonable sort of accommodation for business usage. How would you finance the thing? It seems to me that an understanding of the mortgage market might be helpful. The mortgage market represents a segment of American industry that is probably the most static and one of the most backward. Normally if you approached us to build a house underground we would do one of two things. We would giggle a lot and say to the fellow in the next office—we're both loan officers—I would say, "Hey, come in here and listen to this! This guy wants to build a house underground." And we would talk about it a little bit, asking some silly questions like "What is wrong with houses above ground? Are you afraid of sunlight?" We would have no program to fit you, because we go by a manual published by the government called Housing Programs. We would start with the 110 Program and the 235's which were phased out, and the HR Program, and the 236's and 237AB's, and the 10's and the 1045's, and we don't find underground there at all. But we could do this—we could offer you an interim loan. So you want to build a basement first? We'll give you an interim construction loan at about 12 or 14% to build a basement first. Is that a reasonable alternative? No, it really isn't.

Mortgages are divided into about five different types: commercial, industrial, residential, farms and ranches, and special purpose properties. You would come under two of those with an underground building. Under special purpose properties you would get a higher rate of interest and a shorter term. And the other would be determined by the use to which you would want to put the building. If you understand that, it seems it would be a little easier talking to the loan officer about what you are going to do with the building ultimately.

The mortgage market is a strange market in that its funds come cyclically. About every four and a half years all of the money drains out of our business. We went through a period of disintermediation, which means that money goes away from the traditional financial intermediaries: savings and loans, life insurance companies, banks and trusts, commercial banks, mutual savings banks. It means that there is no money available for home building. In the various recessions we have gone through, six of them since 1947, we find that the home building industry is the first hit and the last helped. So we are a very defenseless industry when it comes to the cycles of the economy. There is not much we can do when the economy bottoms out. The last thing we want to do is to loan money when there is no guarantee of resale.

Now, the mortgage market is a two-tiered market. Suppose that I run a Kroger Store and all of the bread that I sell at the store is sold on credit. A loaf of bread costs a dollar. I sell the bread on time for ten cents now and ten cents a month. You continue to get more bread and it continues to pile up. Eventually all my capital will be gone and I will be collecting notes for the bread I've sold. All my capital will be drained and I will go out of business fairly soon, unless I can find somebody to buy all of those notes for the bread. It is that second tier that buys all of my bread notes. Great Grandfather Bread comes in and says, "Dean, you are operating the Kroger Store pretty well. How many notes on bread do you have?" I respond, "About $16,000's worth." He gives me $16,400 for the notes and I start all over again. I have obtained some dough. The mortgage market works the same way.

You go to a savings and loan and you get a real estate mortgage on a home. They give you $50,000 for thirty years at 8% with two points on the front. You assume that they are collecting their money for thirty years and that they don't get it all back till the end of the thirtieth year. Statistics indicate that you will live in a home only a little less than five years. The mortgage will not last more than eight years, even if you assign it to someone else. So there is a secondary mortgage market much like the Grandfather Bread buyer. It is like the Federal National Mortgage Association, Federal Home Loan Bank Board, or the Government National Mortgage Association that comes in and buys in bulk the mortgages that have been generated in the little savings and loan in Minnesota, Wisconsin, Ohio or Texas. That is how it works.

The way to generate national policy is through the secondary mortgage market level. For example, let us look at some similar kinds of operations. Super highrises: you build a building and then you build living quarters on top of the building. Shopping centers: Highland Park Village, the second shopping center built in America was built in Dallas, Texas in 1928. Mobile homes, community development concepts. These are all new concepts in building much like the new concept of going underground. These are some of the successes. There have been failures as well. A recent failure that we see sliding down a greased tube is the "new town" concept.

How do these operations—super highrises, shopping centers and all of those—work? They are generally backed by large industries. I am a multimillion dollar corporation and wish to borrow one million dollars by pledging a compensating balance of 4 million dollars. To balance one million with a pledge of 4 million is not a stroke of genius in mortgage financing, is it? That is like saying "Can I borrow $10 if I give you $20 to hold until I pay you back?" You would think that pretty good security for a $10 loan. Well, that is how many corporations do their financing through the movement of funds. That does not apply to you or to me, but in terms of national policy, that means something.

Let us look at the mobile home industry, the industry that I find to be the closest to the one of going underground. The mobile home industry has suffered since 1942 in two to three year cycles. A mobile home used to be that thing we towed on the back of the car and parked near lakes. Then they were used for summer homes, then near cities, and we were always trying to do something with the mobile home industry and were shoved away. With the increasing cost of construction, home ownership and financing, we are now recognizing the mobile home and the modular home as a reasonable alternative. It has moved this year into the forefront in financing because the Federal Home Loan Bank Board was convinced by a group of multi-millions of people, such as yourselves, to develop a program that works this way. The Federal Home Loan Bank Board says that during 1975-1976 they will buy one billion dollars worth of loans generated on mobile homes if they are offered at 7 1/2% for 15 years, 20 years and with certain other specifications as to construction. This then makes a real impact on the savings and loan associations. When you go to the savings and loan association and apply for a loan for a mobile home under the program 1045B, which is the mobile home program, they look it up in the book, discover it there, and say, "Well, I'll be darned."

The underground project would work much the same way. You decide as a matter of national policy that it would be reasonable to fund such experiences on an experimental basis. You convince the Federal National Mortgage Association that it would be wise to purchase a certain number of mortgages from savings and loan associations and banks, mortgage bankers and life insurance companies, which then give those people who are generating the loans the liquidity that is needed to even talk to you. That is how it is done. You would never do it with a primary loan. The primary mortgage market is where you would come to my bank and talk to me about a loan. We would laugh and say how strange it is. You have been through the whole thing if you have any interest in underground buildings—you have been through the same reactions I have in trying to get information from people. It is a strange kind of a concept apparently.

Mobile homes, I would guess, were treated the same way in the beginning. Highrise buildings were treated the same way in the beginning. Three years ago we nearly rolled off our chairs talking about mini-warehouses. Why would people want to rent space 10 ft. x 12 ft. high for $10 per month? Isn't that strange? But we look at the size of the average dwelling decreasing in square footage. It decreased by 4% during the period of the sixties, increased in the early seventies, and is going back down again. We find people moving many goods that will be stored. It is a booming business. And it is financed by major industry. How did it get financed? The secondary market was developed first where the loans could be sold. Much like the bread man. He generated all of these credit applications and credit arrangements for bread and finally someone came along and bought them all from him. It is the same way in the mortgage market. In terms of national policy it is easier to put into effect at the secondary mortgage market level. How do you do it? I would presume that you call Gordon Nelson at the Federal National Mortgage Association and ask for an appointment. You start that whole process in much the same way you request a grant from the National Science Foundation. It is a question of requesting, substantiating and response, requesting, substantiating and response, and eventually you are there.

Do we have other programs that are similar? Yes we do. We have the condominium program, the co-operative program, and the air rights program that permits you to buy air space and put in the air space what you want. All of this is through the secondary mortgage market level. The effort of the underground enthusiasts has been basically at the primary mortgage market level, and it will never succeed there. If there is not a market to cover loans, it will not work. There are a few isolated situations, like John Barnard, but generally speaking, I would encourage a strong look at the secondary level.

I have indicated in this paper, some of the things that I would call valleys in our conversation about why the mortgage market is the way it is, how it gets help and how it gets hurt. Now let us talk about some alternative measures that may be developed in addition to the secondary mortgage market area.

If we are going to use earth covered buildings, ecology houses or underground dwellings, how could the government encourage the construction of such? We catalog the variety of ways it can be done so that I can be encouraged to participate in such financial adventures by looking at what similar industries have done over a long period of time. Tax credits or tax incentives are tremendous, and there are the traditional ways of offering a tax incentive. Number two, the availability of a government guarantee. If the government guarantees the loan, then the risk to you, the banker, is substantially reduced. Say that something happens. The greatest potential problem faced by underground houses or businesses is that the market in the future might no longer accept that as an alternative for space. But if I have a government guarantee and go to the banker, at least he is assured that the loan will be a reasonably safe financial venture. The same applies to government insurance, similar to FHA private mortgage insurance. Emphasis must also be placed on the development of a national program encouraging financial intermediaries to invest in such mortgages and to then permit the mortgage to serve as part of the reserve that the bank must hold with the federal reserve. Say the federal reserve requirement is at 16-2/3, represented by certain kinds of securities. We could say at the Federal Reserve Board level, that one of the securities acceptable would be loans not in excess of 1% per year portfolio on underground dwellings. You would encourage the bank to participate at that level as well. Or the last plan would be an agency that would oversee this entire affair—HUD, FEA, or whatever, could obtain the power to float a bond on the capital market to raise money. The method used could be a debenture such as the Fannie Mae debenture or the Federal Home Loan Bank Board debentures, or the latest one we see, the Big Mac debentures. To those who may not be familiar with the shorthand, that is the funding agency that is saving New York City, or is postponing its financial demise for 7 1/2 years. It raises money to then loan to the city. We could develop, through the agencies already existing, a plan to raise money in the capital market. This money could then be used to loan to people, through the banks and the financial institutions, to build homes under the ground.

When contemplating a mortgage loan, there are always two things that we look at. We look at the building and we look at the person. At SMU we say that we look at the appraisal technique and the underwriting possibilities. The appraisal relates to the building and the underwriting relates to the individual. The stronger the building is in and of itself the better the appraisal—if I am building a shell, and I have leased that shell for 99 years to Kroger, and Kroger has signed the lease, then I don't have to be a very strong borrower, do I? I could be a college teacher, because that building arrangement is so strong that it will work. Let us say that I am going to build an underground home for my wife, myself, and our children. It is going to be near SMU. I had better be a pretty strong borrower, as the bank has never seen one of these things before. There is always that relationship between the strength of the building to stand on its own in the event of default, and the strength of the borrower to continue the payments in the event that the building does not turn out the way it is supposed to. So by strengthening the borrower, we are talking about the reduction of risk, compensating balances, and things of that sort. We will be available at SMU on a no-charge basis to assist you in these kinds of problems. We will talk to you on the phone, or correspond with you in any way that you wish. If you want to develop a plan in your own city, we will correspond with you and with the banker that we know in your city, providing him with materials that we have that are available. So if that can be of any help to you, we will be pleased to do it. We can do that either directly at SMU or through the Center for Energy Policy Studies at UTA. It has been my pleasure to be here.

2 references

- earth-sheltered housing plans
- product information
- government or regulatory agencies
- periodicals
- solar energy information
- bibliography

Some additional sources of information will be given in this section. It is inevitable that many worthwhile sources or references will be omitted. If glaring omissions are noted, please contact the Underground Space Center so that the reference lists can be adjusted accordingly for future use. The address to write to is:

The Underground Space Center
11 Mines and Metallurgy Bldg.
221 Church Street S.E.
Minneapolis, Minnesota 55455

earth sheltered housing plans

Some known sources of earth sheltered house designs and plans (other than those included in Part B) are:

Frank Moreland
Center for Energy Policy Studies
University of Texas at Arlington

Solar Earth Energy, Inc.
2020 Brice Road
Reynoldsburg, Ohio 43068

Jay Swayze
Geobuilding Systems
P.O. Box 1556
Hereford, Texas 79045

Obviously, many other people or firms who are presently designing or constructing earth sheltered homes will not have been included in this list or in Part B. House designers and builders can be very difficult to locate other than by word of mouth. To help remedy this situation and to give interested parties a means by which they can contact a local designer or builder, the American Underground Space Association will prepare a list of its members who are involved in earth sheltered housing design. Designers and builders who belong to the Association and who would like their names on this list should contact the Association office. Non-members are invited to join the Association so that they can also be placed on the list. These communications should include: the number of houses completed, the number in progress and some brief comments on the type of houses constructed. Listings will be made available to interested parties who contact the Association.

People requesting these listings in the future should include a stamped addressed envelope and allow 2-3 weeks for a reply. Since this listing will take time to grow, it is unlikely to be worthwhile requesting such a listing until after May—June 1978.

The address of the Association is:

The American Underground Space Association
Dept. of Civil and Mineral Engineering
University of Minnesota
Minneapolis, Minnesota 55455

Registered architects involved in earth sheltered house design can usually be located through a local society of the American Institute of Architects. The Minnesota Society of Architects has a listing of Minnesota architects involved in both solar and underground architecture. The list is periodically updated so it has not been reprinted here. It may be obtained from:

MSAIA
314 Clifton Avenue
Minneapolis, Minn. 55403
(612) 874-8771

There are two major sources for locating building product manufacturers and suppliers. These are:

Sweet's Architectural Catalog File
McGraw-Hill Information Systems Company
Sweet's Division
1221 Avenue of the Americas
New York, NY 10020

Thomas Register of American Manufacturers
Thomas Register Catalog File
Thomas Publishing Company
One Penn Plaza
New York, NY 10001

Both listings should be available in major libraries.

Some special products and sources of information are listed below:

Wood Foundations:

National Forest Products Association
1619 Massachusetts Avenue N.W.
Washington, D.C. 20036

American Wood Preservers Institute
1651 Old Meadow Road
Mclean, Virginia 22101

Concrete and Concrete Block:

Portland Cement Association (PCA)
Old Orchard Road
Skokie, Illinois 60076

Steel Culvert design information:

Armco Steel Corporation
Metal Products Division
Box 800,
Middletown, Ohio

Exterior Rolling Thermal Shutters:

Pease Company
Ever-Straight Division
7100 Dixie Highway
Fairfield, Ohio 45023

Roll-A-Way Shutters, Inc.
393 105 Terrace N.E.
St. Petersburg, Florida 33702

Rolenco
Suite 103
IE Tower
Cedar Rapids, Iowa 52401

government or regulatory agencies

national building code organizations

1. Basic Building Code
   Building Officials and Code Administrators
   International, Inc.
   1313 E. 60th Street, Chicago, Illinois
2. National Building Code
   American Insurance Association
   85 John Street, New York, New York
3. Standard Building Code
   Southern Building Code Congress International
   3617 Eighth Avenue South, Birmingham, Alabama
4. Uniform Building Code
   International Conference of Building Officials
   5360 South Workman Mill Road, Whittier, CA 90601

government agencies

Minnesota Department of Administration
Building Code Division
408 Metro Square
7th and Robert Sts.
St. Paul, MN 55101
(State code administration)

Dept. of Housing and Urban Development (HUD)
Regional Ofc. (ILL., IND., MINN., MICH., OHIO, WISC.)
360 North Michigan Avenue
Chicago, ILL. 60601
Local Office
(financing, programs for solar grants, etc.)

Minnesota Energy Agency
720 American Center Building
150 East Kellogg Blvd.
St. Paul, MN 55101
(energy information center)

Department of Energy
Washington, D.C. 20585
(administration of all federal energy programs)

Minnesota Housing Finance Agency
480 Cedar Street
St. Paul, MN 55101
(financing of low-moderate cost housing in Minn., administration of Minnesota's earth sheltered housing demonstration program)

major I.C.B.O. documents for building design

Uniform Building Code: $18.00*	This regulates all major aspects of building design and construction relating to life, fire and structural safety. It has requirements based on building occupancy, requirements based on fire zones (applies to urban areas) and requirements based on types of construction.
Uniform Building Code Standards $27.50*	This contains the standard specifications referred to in the Uniform Building Code.
Uniform Mechanical Code: $9.80*	This provides regulations for heating, ventilating, air conditioning and refrigeration equipment for buildings.
Uniform Plumbing Code $9.80*	This covers requirements for drain, waste and vent systems, indirect waste systems, sizing and protection of water supplies, private waste disposal systems, gas piping and water heating installations.
Uniform Fire Code: $9.80*	This deals mainly with fire protection measures for specific types of building occupancy or material storage.
Solar Energy Code: $6.00*	This covers the use of solar energy in plumbing and mechanical systems. Included are definitions, material and methods of construction, tanks, collectors, collector sizing and calculations, insulation, tables for degree days and collector tilt factors.
Uniform Housing Code: $2.75*	This provides requirements for housing compatible with the Uniform Building Code. It lists the applicable procedural and space and occupancy requirements but references most of the other provisions such as mechanical and structural requirements.
One and Two Family Dwelling Code: $9.00*	As mentioned above, this is a joint venture between the model code bodies. It does include mechanical and plumbing requirements as well as construction and occupancy.

*Prices at time of writing

periodicals

Listed below are some of the periodicals that regularly carry features on earth sheltered construction or energy conservation.

Underground Space
Journal of the American Underground Space Association
Dept. of Civil and Mineral Engineering
University of Minnesota
MPLS, MN 55455

Popular Science
P.O. Box 2871
Boulder, Colo. 80302

Popular Mechanics
P.O. Box 10064
Des Moines, Iowa 50340

The Mother Earth News
105 Stoney Mountain Road
Hendersonville, N.C. 28739

Alternative Sources of Energy Magazine
Milaca, MN 56352

1. for information on solar energy

a For general information

National Solar Heating & Cooling Information Center
P.O.Box 1607
Rockville, MD 20850
800-523-2929 (Toll Free)

ERDA Technical Information Center
P.O. Box 62
Oak Ridge, TN 37830

Energy Information Center
Minnesota Energy Agency
740 American Center Bldg.
150 E. Kellogg Blvd.
St. Paul, MN 55101

Solar Energy Information
U.S. Dept. of Commerce
National Technical Information Service
5285 Port Royal Road
Springfield, Virginia 22161

b To be put on the Department of Housing and Urban Development's general solar mailing list, write:

Solar Energy Program
HUD
Room 8158
Washington, DC 20410

c The National Bureau of Standards has issued engineering performance criteria for residential solar installations, while NASA has developed interim standards for equipment for nonresidential buildings. A listing of systems approved under these standards is available on request from:

ERDA Technical Information Center
P.O. Box 62
Oak Ridge, TN 37830

d For guidelines in choosing solar equipment:

Office of Consumer Affairs
Public Affairs
Department of Health, Education and Welfare
330 Independence Avenue, S.W.
Washington, DC 20201

e For professional help in design:

ASHRAE (American Society of Heating, Refrigerating, and Air Conditioning Engineers)
Attention: Joseph F. Cuba
Director of Research and Technical Services
234 E. 47th St.
New York, NY 10017

NAHB Research Foundation, Inc. (National Association of Home Builders)
P.O. Box 1627
Rockville, MD 20850

f The International Solar Energy Society has compiled a bibliography of solar material:

ISES Bibliography
12441 Parklawn Drive
Rockville, MD 20852

g For information about wind energy:

Wind Energy Society of America
1700 East Walnut St.
Pasadena, CA 91106

American Wind Energy Association
21243 Grand River
Detroit, MI 48219

h For evaluation of your own energy-related ideas:

Office of Energy-Related Ideas
National Bureau of Standards
Washington, DC 20234

2. for information about building a solar home or installing a solar system in your present home:

American Institute of Architects
1735 New York Ave. N.W.
Washington, DC 20006

National Association of Home Builders
15th and M Streets, N.W.
Washington, DC 20005

Solar Energy Industries Association
1001 Connecticut Ave. NW
Washington, DC 20036

Spetgang, I., and M. Wells, **Your Home's Solar Potential** is available from Edmund Scientific Co., Barrington, NJ, 1976, 60pp. $9.95.

3. for a survey of solar homes in your area:

Shurcliff, William A., **Solar Heated Buildings: A Brief Survey (13th Edition),** 19 Appleton Street, Cambridge, MA 02138, $13.00 postpaid; $14.00 first class mail.

4. solar energy related periodicals:

Alternative Sources of Energy. Alternate Sources of Energy, Inc., Route 2, Box 90A, Milaca MN, Quarterly.

The Mother Earth News. The Mother Earth News, Inc., 105 Stoney Mountain Road, Hendersonville NC 28739, Bimonthly.

Solar Age. Solar Vision, Inc., Rt. 515, Box 288, Vernon NJ 07462 Monthly.

Solar Energy. Pergamon Press Ltd., Maxwell House, Fairview Park, Elmsford NY 10523, Bimonthly.

Solar Energy Digest. CWO-4 W.B. Edmondson, P.O. Box 17776, San Diego CA 92117, Monthly.

Solar Engineering. Solar Engineering Publishers, Inc., 8435 No. Stemmons Freeway, Suite 880, Dallas TX 75247, Monthly.

Solar Heating and Cooling Magazine 20 Community Place, Morristown, NJ, 07960

5. solar directories

Pesko, Carolyn, ed. **Solar Directory.** Ann Arbor: Ann Arbor Science Pub., 1977, 600 pp.

Shurcliff, W.A., **Informal Directory of the Organizations and People Involved in the Solar Heating of Buildings.** W.A. Shurcliff, 19 Appleton St., Cambridge, MA 02138, $5.00 prepaid.

Solar Directory. Environmental Action of Colorado, 2239 E. Colfax Ave., Denver, CO 80206. $20.00/yr.

Solar Energy Directory: A Directory of Domestic and International Firms Involved in Solar Energy. Pheonix: Centerline Corp., 1976, 108pp., $7.50.

Solar Industry Index, 1977. Washington, DC: Solar Energy Industries Assn., 1001 Connecticut Ave. N.W., Suite 632, 20036, $8.00.

6. solar energy related bibliographies:

Energy Research and Technology Abstracts of NSF/RANN Research Reports, Oct. 1970-Dec. 1974. Available from George James, National Science Foundation, ISRU/RANN-Room 1233, Washington, DC 20550. No charge.

Environmental Action. Total Environmental Action, $6.00.

NSF/RANN Solar Energy Heating and Cooling of Buildings. Current Publications Bulletin I;2. NSF/RANN Document Center, National Science Foundation, Washington, DC 20550.

Solar Energy and Wind Power: Selected Bibliography. Springfield, VA: NTIS, 1974, COM-74-1110; $25.00.

Solar Energy Bibliography. Volume 1: Citations; Volume 2: Indexes. Springfield, VA: NTIS, 685 pp., $13.75 paper; 398 pp. $10.75 paper, both $3.00 microfiche.

Solar Heating and Air Conditioning: A Bibliography. Springfield, VA: NTIS, 1974, NTIS/PS-75/34; $25.00 paper or microfiche.

Solar Space Heating and Air Conditioning: A Bibliography with Abstracts. Citations from the NTIS Data Base from 1964-Sept. 1976. Springfield, VA: NTIS, NTIS/PS-76/0727, $25.00.

Solar Thermal Energy Utilization 1957-1974. Albuquerque, NM: Technology Application Center, University of New Mexico 87131, 2 vols., $37.50; or NTIS—PB—238 269, $37.50.

Solar Energy, Other Sources of Energy. I.S.E.S., American Section, 1975, 4pp. no charge.

7. catalogs and surveys of solar energy heating & cooling products:

Superintendent of Documents (catalog of heating & cooling products)
U.S. Government Printing Office
Washington, DC 20402
Order number: ER1.11:ERDA-75. $3.80

Superintendent of Documents (survey of products & services)
U.S. Government Printing Office
Washington, DC 20402
Order number: Y.4SC12:94-1/G $4.60

Solar Energy Industries Association, Inc. (Directory & Buyers guide)
1001 Connecticut Avenue NW
Washington, DC 20036

Solar Industry Index ($8.00 prepaid $2.00 postage & handling)
SEIA
1001 Connecticut Ave. NW
Washington, DC 20036

Buying Solar
U.S. Government Printing Office
Washington, DC 20402
Stock I;: 041-018-00120-4 I6 $1.85/copy

Solar Source Book (700pp; $12.00 prepaid)
SEINAM, Dept. C, P.O. Box 9352
Washington, DC 20005

8. manufacturers and distributors of solar hot water systems:

This list has been reproduced by permission of Solar Heating & Cooling Magazine. The original table, which appeared in the October 1977 issue, includes a tabulation of 14 details for each system.

Alten Corp.
Mountain View, CA 94043

American Heliothermal Corp.
Denver, CO 80237

CSI Solar Systems Div.
Clearwater, FL 33520

Cole Solar Systems Inc.
Austin, TX 78745

Daystar Corp.
Burlington, MA 01803

E & K Service Co.
Bothell, WA 98011

Energy Convertors Inc.
Chattanooga, TN 37406

Falbel Energy System Corp.
Greenwich, CT 06830

Grumman Sunstream
Ronkonkoma, NY 11779

Hitachi Chemical Co. Amer. Ltd.
New York, NY 10022

Inter Technology/Solar Corp.
Warrenton, VA 22186

W.L. Jackson Mfg. Co., Inc.
Chattanooga, TN 37401

Largo Solar Systems Inc.
Plantation, FL 33317

Lennox Industries
Marshalltown, IO 50158

Mar-Scot Pac Systems
Miami, FL 33157

Mor-Flo Industries Inc.
Cleveland, OH 44128

National Solar Supply
Atlanta, GA 30318

Natl. Energy Sys. Corp., (NESCO)
Birmingham, AL 35201

PPG Industries
Pittsburgh, PA 15222

Raypak Inc.
Westlake Village, CA 91359

Revere Copper & Brass Inc.
Rome, NY 13440

Scientific Atlanta
Atlanta, GA 30340

Simons Solar Environmental Inc.
Mechanicsburg, PA 17055

Sol-Therm Corp.
New York, NY 10011

Solar Development Inc.
West Palm Beach, FL 33407

Solar Energy Products Inc.
Gainesville, FL 32601

Solar Energy Systems, Inc.
Cherry Hill, NJ 07003

Solar Enterprises Inc.
Arlington, TX 76012

Solar Industries of Fla.
Jacksonville, FL 32208

Solar Innovations
Lakeland, FL 33801

Solar Kinetics Corp.
West Hartford, CT 06117

Solar Res. Div. of Refig. Res. Inc.
Brighton, MI 48116

Solar Systems Inc.
Racine, WI 53403

Solaron Corp.
Denver, CO 80222

Sunearth Solar Products Corp.
Green Lane, PA 18054

Sunworks
New Haven, CT 06508

Thomason Solar Homes Inc.
Washington, DC 20022

Ying Mfg. Corp.
Gardena CA 90249

Energy Information

Minnesota Energy Use Trends 1957-73 February 1975, 25 pp., Minnesota Energy Agency Research Division
An assessment of trends in the consumption of all forms of energy and an analysis of the social, economic, and environmental consequences of these trends.

Alternative Energy Sources Conference Summary Update, Technology and Applications for Minnesota, April 27-28, 1976, Bloomington, MN, 57 pages
Contains 2-3 page articles on different types of alternative energy sources and their potential for use in Minnesota.

Minnesota Alternate Energy Research and Development Policy Formulation Project Minnesota Energy Agency Research Division, 1977, V.I Draft executive summary & recommendations V.2 Solar I, Active V.3 Solar Passive V.4 Urban Solid Waste pts. A-B
A series of reports on alternate energy sources and Possibilities. Contains detailed information on many aspects of the specific topics listed.

Energy Use and Conservation in the Residential Sector: A Regional Analysis
Stephen H. Dole R-1641-NSF, June 1975 (Rand Corporation).
A detailed study of residential energy consumption in different sectors of the country.

Energy Conservation in Building: its Foundation, Costs and Acceptance Paul Reece Achenbach, Energy Conservation Conf., Univ. of Minnesota, 1973, 38 pages
This paper discusses energy usage and the need for improved thermal design in residential buildings. Does not deal specifically with underground buildings.

Conferences and Symposia

Energy Use Management Rocco A. Fazzolare and Craig B. Smith, ed., Pergamon Press, Inc.1977. Four volume proceedings of the International Conference on Energy Use Management held October 24-28, 1977, in Tuscon, Arizona
These proceedings are a massive source of information on the state-of-the-art in energy conservation. Papers cover a very wide range of applications of energy use management.

Alternatives in Energy Conservation: The use of Earth Covered Buildings NSF/RA-760006 Proceedings and Notes of Conference held in Fort Worth, Texas, July 9-12, 1975, $3.25. Available from U.S. Govt. Printing Office, Washington, DC 20402 Stock No. 038-000-00286-4
This book contains a large collection of papers on the use of underground space. Several underground projects are reviewed as well as possibilities and problems discussed. Contains an extensive bibliography.

Legal, Economics and Energy Considerations in the use of Underground Space National Academy of Sciences, Washington, DC, 1974
A collection of papers presented at the June 1973 Engineering Foundation Conference. The papers deal with the title topics.

Need for National Policy for the use of Underground Space Engineering Foundation Conference. Berwick Academy, South Berwick, Maine, June 25-29, 1973, 232 pages
This book contains a collection of papers presented at the Conference on: identifying competitive claims for underground space, how existing policies developed, current and anticipated problems, and the development of future policies.

Proceedings of the Symposium on the Development and Utilization of Underground Space University of Missouri, Kansas City, 1975
Mainly concerned with the use of deep underground space for commercial and industrial facilities.

Going Under to Stay on Top Proceedings
From a two day conference held October 14-15, 1977 in Minneapolis, Minnesota and co-sponsored by the American Underground-Space Association and the Science Museum of Minnesota. Proceedings will be available shortly. Contact either the AUA or the Science Museum for further details.

General Articles

Going Under to Stay on Top
Charles Fairhurst, Underground Space Vol. I, No. 2, July-Aug., 1976, pp. 71-86
This paper reviews some of the possibilities of the use of underground space in a wide range of applications.

Energy Conservation by Building Underground Thomas P. Bligh, Underground Space, Vol. I, No. I, May/June, 1976, pp. 19-33

Conservation of Energy by Use of Underground Space Thomas P. Bligh and Richard Hamburger in: Legal, economic and energy considerations in the use of underground space National Academy of Sciences, 1974, 15 pages, Bibliography
Discusses the energy savings of underground buildings and also additional advantages when manufacturing plants are located underground.

Non-Traditional Military Uses of Underground Space Lloyd S. Jones, Underground Space, Vol. 2, No.3, Nov., 1977
Discusses the advantages of military uses of underground space for bulk fuel storage, ammunition storage, medical facilities, warehousing, administrative space, special purposes and housing and quarters

Is it Time to go Underground? Lt. Lloyd Harrison, Jr. The Navy Civil Engineer, Fall 1973, 2 pages
Discusses the advantages of underground construction.

Why I Went Underground Malcolm Wells, The Futurist, February 1976, pp. 21-24
Malcolm Wells discusses the reasons for turning to underground architecture and what added advantages he discovered. The article contains illustrations and floor plans of his underground office.

What Your Home Will be Like in Fifty Years Malcolm Wells, Mainlines, September 1974.

Confessions of a Gentle Architect Malcolm Wells, Environmental Quality, July 1973.

Conservation Architecture Malcolm B. Wells, Architecture & Engineering News, September 1969, p. 27.

Down Under, Down Under, ... or How Not to Build Underground Malcolm Wells, Progressive Architecture, March 1968, 2 pages
Short essay on ugly usage of underground housing in the Australian outback.

Nowhere to go but Down Malcolm Wells, Prog. Architecture, February 1965, pp. 174-179

Energy Essays Edmund Scientific Company, Malcolm Wells, 1976, 70 pp. $5.95
Illustrated essays on solar energy, nuclear power, air, water, food, waste-management and energy smugness.

Geotecture Royce LaNier, Library of Congress, Catalogue No. 76-139958, 1970, 72 pages.
A book outlining the history and potential of underground space use. (with bibliography).

Geotecture, Concept, Design, Construction and Economy of Geospace - The Creation of Subterranean Accommodation Patrick Horsbrugh, Univ. of Notre Dame, 1973.

Underground Housing R.F. Dempewolff, Science Digest, Nov. 1975, pp. 40-53

Underground Construction Buildings, Jan. 1977, 6 pages
Discusses the advantages of underground construction in terms of energy saving, land conservation an environmental dividends. It also compares initial and long term cost.

Underground Architecture: What Lies Ahead May be Beneath Us Roy Mason, The Futurist, Feb. 1976, pp. 16-20
This article discusses the advantages of underground construction and reviews some existing and proposed underground structures.

Saving by Going Underground AIA Journal, Feb. 1974

Down to Earth Housing Wybe J. Van der Meer, Solar Age, Sept. 1976

The Earth's the Ceiling James Gorman, The Sciences, March/April 1976

Energy Conservation Forces a Revolution in Buildings Engineering News Record, N-703, 1975, 12 pages.
Discusses the need to conserve energy and describes ways of doing it in existing and new buildings.

Houses: The Architect Speaks to Man's Needs P/A, May 1964, 5 pages
An essay on the need for distinctive houses for distinctive people including earth form houses.

Land Use Planning

Earth Covered Habitat - An Alternative Future Frank L. Moreland, Underground Space, Vol. I, No. 4, Aug. 1977, pp. 295-307.

Examines the viability of earth covered housing from a consumer perspective.

Planning the Underground Uses Donald F. Hagman, Legal Economic, and Energy Considerations in the use of Underground Space. National Academy of Sciences, 1974, pp. 52-68.
This paper discusses the planning and policy considerations for underground space in general.

Earth Shelter Architecture and Land use Policy Royce LaNier and Frank L. Moreland, Underground Space, Vol. I, No. 4, Aug. 1977, I page editorial
This article discusses the need to change our land use habits and the laws governing them in order to significantly reduce energy consumption.

Assessing Environmental Impact of Earth Covered Buildings Royce LaNier, Underground Space, Vol. I. No. 4, Pergamon Press, Aug. 1977, pp. 309-315.
This paper looks at the environmental impact of earth covered buildings and at the potential for minimizing the environmental impact of urbanization by greater use of this technique.

Subterranean Urban Systems Gunnar Birkerts, 1974, Industrial Development Division, Institute of Science and Technology, University of Michigan.

Liberating Land: A Blueprint for Urban Growth G. Birkerts, Progressive Architecture, Mar. 1973

Planning of Subsurface Use Birger Jansson & Torbjorn Winguist, D7:1977, 170 pages
Detailed report outlining the urban planning process for wide use of underground space. Based on Swedish experience in underground construction.

I: Potential use of Underground Space Minneapolis Campus, Univ. of Minnesota, Prepared by: Department of Civil and Mineral Engineering, U of M, Feb. 1975, 75 pages, $7.00
This report is concerned with the future use of underground space by the U of M and outlines the benefits of such use. Its objectives are to include the use of underground space as an alternative for future projects and to outline factors important in underground planning.

II: Preliminary Design Information for Underground Space Minneapolis Campus, Univ. of Minnesota, Prepared by: Department of Civil and Mineral Engineering. U of M, Aug. 1975, 98 pages, $7.00.

Architectural Design

Underground Designs Malcolm Wells, 1977, 87 pages, $6.00. Available from Malcolm Wells, P.O. Box 1149, Brewster, MA 02631 or some bookstores.
This book contains plans and illustrations of Wells' many underground designs, (18 home designs). Not all of the designs were actually built but they provide a valuable idea book of solutions to the problems of designing underground buildings. The book also contains short sections dealing with selecting a site, choice of structure, building code problems, waterproofing, insulation and landscaping.

Earth Integrated Architecture 1975, Edited by and available from, James W. Scalise, College of Architecture, Arizona State University, Temple, AZ
A collection of student research and design projects on earth integrated architecture. It contains detailed information on some of the fundamental determinants of earth integrated architecture, and also contains a section of examples and a bibliography. It places a special emphasis on the climate and conditions of the southwest.

Underground Architecture Malcolm Wells, CoEvolution Quarterly, Fall 1976, Box 428, Sausalito, California 94965
Malcolm Wells describes the advantages of living underground. He also describes his underground home and office. Several design tips included.

The Energy Factor - A Dimension of Design David Bennett & Thomas P. Bligh, Underground Space, Vol. I No. 4, Aug. 1977, pp. 325-332
This paper describes the approach to achieving a balance between technical efficiency (energy use) and human needs as it was applied to the design process of Williamson Hall, Univ. of Minnesota.

Terratecture J.W. Scalise, 1974 published by Arizona State University

The Architectural Underground Part I Kenneth Labs, Underground Space, Vol. I No. I May-June 1976, pp. 1-8.
This article is concerned with the use of underground space for architectural application. Part I examines the primary themes around which historical underground applications

usually revolve.

The Architectural Underground Part II Kenneth Labs, Underground Space Vol. I No. 2, July-Aug. 1976, pp. 135-156
Part II examines the modern forms and functions of underground construction with the intention of summarizing our current state of practice.

The Architectural Use of Underground Space: Issues and Applications Kenneth B. Labs, Master of Architecture Thesis, Washington Univ., St. Louis, MO, May 1975
A detailed summary of many of the considerations in designing earth sheltered structures.

Subsurface and Underground Housing in Hot Arid Lands A technical report, U.S. Army Research Office, Durham, North Carolina, 1964.

Underground - The Prairie Sod House Returns Architecture Minnesota, Sept.-Oct. 1977, pp. 24-31.
Contains reviews of a proposed Fort Snelling Center and Williamson Hall, U. of M.. It also discusses the advantages of underground architecture.

Principles of Natural Lighting J.A. Lynes, 212 pages, 106 illus., 1968, $15.00, Applied Science Publishers Ltd., 22 Rippleside Commerical Estate, Rippleroad, Barking, Essex, England

Earth Sheltered House Reviews

Solaria Edmund Scientific Company, Bob and Nancy Homan, Dr. Harry Thomason, Malcolm Wells, April 1974, 67 pp. with 10 detailed plans of Homan House. $24.95, from Edmund Sci.
Contains descriptions and plans of an earth covered solar house, Solaria, (see part B).

Underground Houses V.E. Smay, Popular Science, p. 84, April 77
This article reviews several existing underground houses and discusses their merits, (photographs and illustrations).

Your Next House Could Have a Grass Roof R.F. Dempewolff, Popular Mechanics, p. 78, March 1977.
This article describes the different types of underground architecture. It reviews some existing and proposed underground houses and discusses their advantages, (photographs & illustrations).

Winston House, Lyme, New Hampshire, Designer and Contractor Don Metz Architectural Record Houses of 1974, pp. 52-53, Progressive architecture, May 1975, (See Part B)

My Cave Andy Davis, Underground Space, Vol. 2, No. 3, Nov. 1977.
A description of Andy Davis' earth sheltered house in Armington, Illinois. Plans, data and photographs included (See Part B also)

Plowboy Interview Andy Davis' Cave, Mother Earth News, July/Aug 1977.
An interview with Andy Davis discussing the construction and performance of his earth sheltered home in Illinois. Plans, data and photographs included. (See Part B also).

Underground Living in This Ecology House Saves Energy, Cuts Building Cost, Preserves the Environment V. Elaine Smay, Popular Science, June 1974, pp. 88-90
A review of John Barnard's ecology house. (See Part B)

Living History Farms, The Farmhouse of Today and Tomorrow Ames Design Collaborative, Ray D. Crites, Sept. 28, 1976, 17 pages.
A description of an experimental farmhouse at Urbandale, Iowa, that was designed to have normal liveability, be energy conservative and use materials that are plentiful and not energy intensive to refine. Contains design, structure, and environmental control system information and illustrations.

Going Underground: House of Tommorrow Bonnie Speer, Orbit Magazine, The Sunday Oklahoman, April 10, 1977.
A review of several underground homes in Oklahoma with reactions from their owners. Illustrated

An Instinctive Home Progressive Architecture, May 1964, p. 187, 8 pages
A review of a house designed by Jacques Couelle, located in Castellaras-le-Neuf, Alpes-Maritimes, France. It has a cave-like form carved from the rocks. Many illustrations.

The Self-Heating, Self-Cooling House Wendell Thomas, Mother Earth News Reprint, #39
A review of two underground houses (Sunnycrest & Sunnycave), designed and built by Wendell Thomas, in the mountains of western North Carolina. Describes the design of the houses and the provisions for heating and cooling.

The Beale Solar-Heated Subterranean Guest House Mother Earth News, May/June 1977, p. 80-81.
This article describes a passive solar heated guest house in Athens, Ohio.

Go Underground Young Man, Says John Strickler Seattle Post-Intelligencer Northwest, Sun., March 6, 1977, by Linda Lewis, 3 pages.
A review of John Strickler's house which is located on Camano Island.

30 Energy Efficient Houses ... You can Build Alex Wade and Neal Ewenstein, Rodale Press, Emmaus, PA, 1977, 316 pages, $8.95
Descriptions with floor plans of 30 houses, most of them owner built, which were designed to be space and energy efficient. Short section on earth sheltered and passive solar design.

Energy Use Factors in Houses & Buildings

ASHRAE Handbook of Fundamentals American Society of Heating, Refrigerating and Air-Conditioning Engineers, Inc., 345 East 47th Street, New York, NY 10017.
This multi-volume book is the standard reference for heating and cooling calculations in buildings and the design of HVAC systems.

Heat Loss Through Basement Walls and Floors F.C. Houghten, D.I. Taimuty, Carl Gutberlet and C.J. Brown, Pittsburgh, PA, pp. 369-384, Transactions American Society of Heating and Ventilating Engineers, No. 1213, 1942.
Technical paper containing the results of research on heat loss through basement walls and floors. Excellent basic data from a well instrumented test basement.

Heat Losses From House Basements Canadian Buildings, No. 10, Oct. 1969, 4 pages.
Detailed report of technical study of heat loss from basement walls and floor at different depths beneath the surface. Also a table of heat loss from a typical bungalow with both above and below grade portions and with different amounts of insulation.

Temperature and Heat Loss Characteristics of Concrete Floors Laid on the Ground H.D. Bareither, A.M. Fleming, and B.C. Alberty, University of Illinois, Small Homes Council, PB 93920, 1948.

Measurements of Heat Losses from Slab Floors R.S. Dill, W.C. Robinson and H.E. Robinson, National Bureau of Standards Report BMS 103, March 10.
This paper presents the results of Bureau tests of eight floors. It gives quantitative information that may be used for estimating heat losses through floors and a means by which typical floors can be compared on the basis of heat transfer properties.

Heating and Air Conditioning of Underground Installations Dept. of the Army Technical Manual, Aug. 1965. TMS-8554, Washington, DC.
This manual presents criteria, and engineering information and procedures for the design of heating and air conditioning systems for underground installations within limits satisfactory for human occupancy and the preservation of equipment, supplies and materials.

Air Conditioning is Easier Underground Heating and Ventilating Engineer, Jan. 1963.

Numerical Analyses of the Thermal Environment of Occupied Underground Spaces with Finite Cover using a Digital Computer T. Kusuda, R. Achenbach, ASHRAE Transactions, Vo. 69, No. 1853, 1963, 14 pages.
This paper describes a method for calculating the temperature, humidity conditions and heat transfer in experimental underground fallout shelters for periods of occupancy up to two weeks.

A Study of Thermal Environment in Underground Survival Shelters Using an Electronic Analog Computer E.E. Drucker, J.T. Haines, ASHRAE Transactions, Vol. 70, No. 1857, 1965, 14 pages.
Results of detailed technical study of an underground concrete shelter designed for 50 people. The object was to predict the thermal environment of the shelter as a function of ventilation rate, physical properties of the earth and air conditioning.

Thermal Inertia - The Neglected Concept M.J. Catani and S.E. Goodwin, The Construction Specifier, May 1977, 5 pages.
A detailed explanation of thermal inertia and its benefits. It deals only with concrete and metal buildings but the idea is applicable to earth shelter.

Heavy Building Envelopes and Dynamic Thermal Response M.J. Catani and S.E. Goodwin, ACI Journal, Feb. 1976,

pp. 83-86.
This paper contains a detailed discussion on the design of building envelopes in terms of material selection, heat transfer, and the effect of mass on thermal response.

Study Credits Use of Mass to Save Energy in Heating and Cooling of Buildings Air Conditioning, Heating & Refrigeration News, Nov. 29, 1976.
A short article describing a study relating mass to thermal performance of a building.

Field Laboratory for Heating Studies D.B. Anderson, G.A. Erickson, R.C. Jordon and R.R. Leonard, ASHRAE Transactions, 1961, 4 pages.
Detailed technical report of heat loss, temperature and moisture studies of two test houses in Stillwater, MN. The test houses are above ground and identical except for insulation.

Infiltration Measurements in Two Research Houses R.C. Jordan, G.A. Erickson and R.R. Leonard, ASHRAE Transactions, 1963, 6 pages.
Detailed technical report of heat loss, temperature and moisture studies of two test houses in Stillwater, MN. The test houses are above ground and identical except for insulation.

Skylight Energy Performance Center for Industrial and Institutional Development, Univ. of New Hampshire, Durham, New Hampshire, Oct. 1975, 16 pages.
An evaluation of the effect of skylights on building energy requirements.

Systems Analysis for Skylight Energy Performance Center for Industrial and Institutional Development, Univ. of New Hampshire, Durham, New Hampshire, Oct. 1975, 23 pages.
A procedure for the systems analysis of the effect of skylights on building lighting, cooling and heating systems and energy requirements.

Energy Conservation in Building Design The American Institute of Architects, May 1974, reprinted June 1976, 156 pages.
This report discusses some of the ways energy can be saved in building designs. It deals with energy use in buildings, policy, physical design alternatives and alternative power sources.

Energy Conservation by Subsurface Construction Drs. Ernest Eckert, Thomas P. Bligh, and Emil Pfender, NSF RA-76043I, Dept. of Mech. Eng., Univ. of Minnesota.
A progress report containing details of the instrumentation program for the study of heat loss in large underground buildings.

Soil Properties & Temperatures

Earth Temperature Beneath Five Different Surfaces T. Kusuda, National Bureau of Standards, Report 10373, Feb. 1971.
This is a final report covering an experimental effort of measuring, processing and analyzing earth temperature data under five different surfaces. The five surfaces were blacktop, a white painted surface, bare ground, short grass, and long grass.

Thermal Properties of Soils M.S. Kersten, U of M, Minnesota Engineering Experiment Station Bulletin, No. 28, Vol LII, No. 21, June 1, 1949, 94 pages.
Detailed technical report on thermal properties of soils.

Earth Temperature and Thermal Diffusivity at Selected Stations in the United States P.R. Achenbach, T. Kusuda, ASHRAE Transactions, Vol. 71, No. 1914, 1965, pp. 61-75.
Detailed information and calculations of earth temperature and thermal diffusivity at selected stations in the United States.

Collins' Well Water Isotherms Ground temperature distribution in the United States, 1 page.
A chart of ground temperature distribution in the United States.

Soil Temperature in the United States E.M. Fitton and C.F. Brooks, Monthly Weather Review, Vol. 59, Jan. 1931.

Soil Temperature B.J. Fluker, Soil Science Journal, Jul.-Dec. 1958.

Soil Thermal Properties and Annotated Bibliography A. Peters, The Franklin Institute, Philadelphia, PA, 1962.

Ground Temperature Jen-hu-Chang, Vol. I & II, Blue Hill Metrological Observatory, Harvard University, Milton, MA, June 1958.

Thermal Soil Mechanics A.R. Jurnickis, Rutgers University Press, 1966.

Ventilation & HVAC Equipment

Air Quality Maintenance in Underground Buildings Robert

Thompson, Underground Space, Vol. I, No. 4, Aug. 1977, pp. 355-364.
This article discusses the application of heating, ventilating and air conditioning systems in below grade buildings. The article deals mainly with medium size buildings such as schools, office buildings and manufacturing facilities.

Symposium on Mechanical Ventilation, its Application and Control for Energy Conservation DA-76-20, ASHRAE Transactions, Vol. 82, Part I, 1976, pp. 1141-1181.
Detailed technical paper that discusses the impact of ASHRAE ventilation standard 63-73 on energy use, mechanical ventilation system design features and control of ventilation air in energy efficient systems.

Recirculation of Exhaust Air U.S. Dept. of Health, Education and Welfare. Division of Physical Sciences and Engineering, Feb. 1976, 266 pages.
The topics covered in this report include threshold limit values for toxic substances, cost-effectiveness and performance standards. Various techniques for monitoring contaminants are discussed and specific design suggestions given.

Measurement of Natural Draft N.W. Isenberg, OCDOS6264, Department of Architectural Engineering, the Penn State University, University Park, PA, Dec. 1963.

Feasibility of Low Cost Ventilation Techniques T. Hori.

Summary of Natural Ventilation of Underground Shelter OCDO562114, MRD Division, General American Transportation Corporation, Niles, Ohio, June 1964.

The Feasibility of Augmenting Below Grade Shelter Habitability with Conditioned Air John Jr. Everetts, Department of Architectural Engineering, Institute for Building Research, Penn. State University, University Park, Penn.

Heat Pump's Ice Storage Cuts Heating-Cooling Costs Architectural Record, Nov. 1976, p. 133, 3 pages.
Describes an experimental heating and cooling system that is installed in a demonstration home in Tennessee. The system utilizes ice making for heat pump heating in the winter and free cooling in the summer.

House Design & Construction

The Owner Built Home Ken Kern, 1972, 1975
Contains detailed suggestions on how to build an inexpensive functional home yourself. Includes areas such as: site and climate, materials and skills, form and function, and design and structure.

From the Ground Up John N. Cole and Charles Wing, Atlantic Monthly Press, Little Brown and Company, 1976, 244 pages.
Information and illustrations intended to provide guidance on designing and building an energy efficient shelter. Not specifically on earth sheltered houses.

Structure, Waterproofing & Insulation

Foundations on Swelling or Shrinking Subsoils J.J. Hamilton, Canadian Building Digest, March 1977, pp 184-1 to 184-4
Discusses problems of foundations on swelling or shrinking subsoil and how to deal with those problems.

Concrete Masonry Basements 13 pages, Available from: Portland Cement Association, 33 West Grand Ave., Chicago, IL 60610.
Contains information on how to build concrete masonry basements.

Recommended Practice for Building Watertight Basements with Concrete 1972, 5 pages, Available from: Portland Cement Association, 33 West Grand Ave., Chicago, IL 60610.
Gives information on how to improve the watertightness of concrete basements. Includes construction details, mortar & concrete mix design and general precautions.

Underground Concrete Mechanical Engineering 92, October 1970, p. 45.

Evaluation of Buried Corrugated Steel Arched Structures and Associated Components G.H. Albright, U.S. Department of Commerce, National Bureau of Standards, Washington, DC, Feb. 1961.

All Weather Wood Foundation System Design, Fabrication, Installation Manual National Forest Products Association, 1619 Massachusetts Avenue, N.W.,Washington, DC 20036, 74 pages.
This manual is in three parts. Part I deals with structural design and details for architects, engineers and draftsmen. Part II covers quality fabrication of the foundation. Part III is concerned with method of installation.

The All-Weather Wood Foundation: Why, What and How 35 pages, Available from: The American Plywood Association.
Discusses the advantages of wood foundations and gives illustrations and detailed description of how wood foundations are constructed.

The All-Weather Wood Foundation System Basic requirements, TSD Technical Report No. 7, (National Forest Products Association).
This report sets forth the minimum basic requirements for the all-weather wood foundation system. Its appendices demonstrate the application of those requirements.

CCA - Treated Wood Foundations, A Study of Permanence, Effectiveness, Durability and Environmental Considerations R.D. Arsenault, American Wood Reservers, Association, 1975, 23 pages.
A detailed technical report. Contains many tabulated results of the study on the permanence of the preservative treatment, the leachability of the chemical and any possible toxicity problems.

Waterproofing Buildings Below Grade G.W. Gill, Civil Engineering 29, Jan. 1959, pp. 3-5.

Subgrade Waterproofing Series in Building Research, Nov.-Dec. 1964.

Evaluation of Bentonite Clay for Waterproofing Foundation Walls Below Grade Stanley M. Kanarowski, Army Construction Engineering Research Laboratory, Champaign, IL AD-AOII 180, May 1975, 48 pages. Distributed by: National Technical Information Service.
A report of a study designed to investigate and evaluate bentonite clay as a waterproofing material for exterior below-grade masonry and concrete foundation walls.

New Basement Wall Designs for Below-Grade Living Space A. Elmroth and I. Hoglund, Canada Institute for Scientific and Technical Information, Ottawa, 44 pages, Technical translation, No. 217.
This is a technical report which describes field tests of a new damp-proofing technique in which mineral fiber insulation is utilized on the outside of basement walls to provide both a drainage layer and thermal insulation. Research was carried out in Sweden based on laboratory research in Norway.

New Method of Drainage of Basement Walls Knut I. Edvardsen, National Research Council of Canada, Technical Translation 1603, Ottawa, 1972, 15 pages.
This is a translation of a study for designing and installing peripheral drainage systems around subgrade construction, carried out by the Norwegian Building Research Institute.

Synthetic Rubber Canal Lining M.E. Hickey, Engineering Research Center, Bureau of Reclamation REC-ERC-71-22, April 1972, 34 pages.
Contains a report on laboratory and field investigations of synthetic rubber sheeting for canal lining.

Nylon-Supported and Unsupported Rubber Linings Vernon L. Kuehn, Engineering and Research Center, Bureau of Reclamation, REC-ERC-71-39, Oct. 1971, 20 pages.
A progress report on laboratory investigations of physical properties of nylon-supported and unsupported rubber linings for waterproofing. The report mainly deals with their use as canal and reservoir liners but is applicable to all waterproofing applications.

Soil Considerations in Subgrade Waterproofing Building Research, Nov.-Dec. 1964, pp. 39-41.
Detailed information on drainage properties of soil.

Damp Diffusion and Buildings Karl Seiffert, translated by A.B. Phillips and F.H. Turner. 209 pages, 44 illustrations, 1970, $21.00, Available from: Applied Science Publishers, Ltd.

Condensation in Buildings D.J. Croone and A.F.C. Sherratt, 271 pages, 52 illustrations, 1972, $25.00, Applied Science Publishers, Ltd.

Thermal Insulation S.D. Probert and D.R. Hub, 121 pages, 62 illustrations, 1968, $12.00, Applied Science Publishers, Ltd.

Laboratory and Field Investigations of Moisture Absorption and its Effects on Thermal Performance for Various Insulations F.J. Dechow and K.A. Epstein, prepared for A.S.T.M. symposium on advances in heat transmission measurements, Philadelphia, 1977.
A detailed study of the moisture performance characteristics of various thermal insulations under service conditions. It discusses freeze-thaw cycling, water absorption and water vapor transmission and their effect on the thermal efficiency of insulation.

Solar Design

Designing and Building a Solar House: Your Place in the Sun Donald Watson, Garden Way Publishing, Charlotte, Vermont, 05445, May 1977, 284 pages, $8.95.
This book contains detailed discussion covering the principles of solar heating, passive systems, active systems, how the parts work together, designing for energy conservation, solar design in northern climates and finally tips for building a solar home.

The Solar Home Book - Heating, Cooling and Designing with the Sun Bruce Anderson with Michael Riordan, Cheshire Books, Harrisville, New Hampshire, 1976, $8.50, 297 pages.
This book discusses in detail the fundamentals of solar heating, direct solar heating, different technology approaches, indirect solar energy systems and do-it-yourself methods.

Solar Dwelling Design Concepts U.S. Dept. of Housing and Urban Development by the AIA Research Corporation, Washington, DC, Contract IAA H-5574, 146 pages, May 1967.
A detailed technical report on existing solar dwellings and systems, a list of design considerations, and numerous dwelling and site design concepts.

How to use Solar Energy in Your Home and Business Ted Lucas, Ward Ritchie, Press, Pasadena, CA 1977, 315 pages, $7.95.
This book contains detailed information on different solar systems to be used for different heating and cooling needs. It also discusses passive solar heat and conservation techniques.

An Inexpensive Economical Solar Heating System for Homes J.W. Alfred, J.M. Shinn, Jr., E. Kirby, and S.R. Barringer, Langley Research Center, Hampton, VA 23665, July 1976, 56 pages, NASA TM-X-3294.
This report describes a low-cost solar home heating system to supplement present warm-air heating systems.

How to Trap Solar Heat with your Windows Edward Allen, Popular Science, p. 108, Feb. 1975

Lee Porter Butler: He Builds Houses That Never Cool Off Hudson Home Guide, Home Building and Remodeling, pp. 62-64.
This article describes Lee Porter Butler's design of passive solar energy houses. Does not deal with underground houses specifically.

Solar Age Catalog Solar Vision, Inc., 1977, 232 pages, $8.50.
A guide to solar energy knowledge and materials. 250 product descriptions with feature by feature comparisons, state by state listings of solar Professionals, and basic articles by solar experts.

Landscape & Plantings

Plants, People and Environmental Quality Gary O. Robinette, Government Printing Office, 1972.
An excellent collection of information of the effect of plants on the surrounding environment. Examples of temperatures, wind, noise and ventilation modification by planting are given.

Landscape Architecture Above Buildings Thomas E. Wirth, Underground Space, Vol. I., No. 4, Aug. 1977, pp. 333-346.
A detailed discussion of structure, waterproofing, planting materials, mechanical systems, cost and coordination which will influence the landscape development of rooftop spaces.

Roofscape Paul M. Friedberg, Architectural and Engineering News, Sept. 1969.

Buildings as Landscape: Five Current Projects by William Morgan William Morgan, Architectural Record, Sept. 1972.

The Climate Near the Ground Rudolf Geiger, Harvard University Press, 1965.

Psychological

On the Psychology of Earth Covered Buildings Paul B. Paulus, Underground Space, Vol. I, No. 2, July-Aug. 1976, pp. 127-130.
A theoretical analysis of the psychological processes which may be important in understanding the potential psychological impact of earth covered buildings.

Studies of Children in an Underground School Frank W. Lutz, Underground Space, Vol. I, No. 2, July-Aug. 1976, pp. 131-134.
This paper outlines two psychological studies carried out on children who attended the Abo elementary school in Artesia, NM, (actual studies listed below). Also was the first underground school of its kind, designed to also act as a fallout shelter.

Abo Revisited Frank W. Lutz, P.D. Lynch and S.B. Lutz,

Defense Civil Preparedness Agency, ContractDAHC20-72-C-0115, June 1972.

Interim Report of the Abo Project Frank W. Lutz and B. Susan, Office of Health, Education and Welfare, Contract OE-3-99-003, Jan. 1964.

Windowless Buildings Tight spaces: Hard architecture and How to Humanize it. Robert Sommer, Englewood Cliffs, NJ, Prentice Hall, Inc., 1974, pp. 114-119.
This article contains interviews with people working in windowless buildings and discusses their feelings about such buildings.

The Availability of Sunshine and Human Requirements for Sunlight in Buildings J. Longmore and E. Ne'eman, Journal of Architectural Research, Vol. 3, No. 2, May 1974, 6 pages.
Detailed information about sunshine duration in London. Discussion of peoples' feelings for sunshine in indoor spaces.

Windows and People: A Literature Survey, Psychological Reaction to Environments with and Without Windows June 1975, Belinda Lowenhaupt Collins, ed., National Bureau of Standards, Gaithersburg, MD.

Sunlight in Buildings Journal of Architectural Research, Vol. 3, No. 2, May 1974.

Public Policy Issues

Equity Requirements of Earth Covered Buildings and Instruments of Remedy Dean A. Manson, Underground Space, Vol. 1, No. 2, July-Aug 1976, pp. 123-125.
Discusses the problems of obtaining mortgages and loans for underground houses, (Reprinted in Appendix F).

Legal Aspects of the Use of the Underground Dan A. Tarlock, Legal, Economic, and Energy Considerations in the use of Underground Space, National Academy of Sciences, Washington, DC, 1974, 11 pages.
Discusses legal aspects of the use of underground space and in particular, ownership of the underground. Explores the need for incentives and restrictions on underground space use.

Insuring Risks Underground - Some General Considerations E.M. DeSaventhem, Underground Space, Vol. 2, No. 1, Sept. 1977, pp. 19-25.
This paper describes how insurers' attitudes will vary between different classes of underground construction. Applies mostly to large scale commercial ventures.

Social Assessment - A Means of Evaluating the Social and Economic Interactions Between Society and Underground Technology Eugene L. Foster, Underground Space, Vol. 1., No. 1, May-June 1976, pp. 61-63
Discusses the need to make a social assessment of individual underground projects rather than a continuous promotion of them.

Late Additions

The Woodburners Encyclopedia Jay Shelton, Published by Vermont Crossroads Press, Waitsfield, Vermont, Feb. 1977

Energy Conservation in Buildings: The Shape of the System Gordon H. Reigstad, President of Nilcon Minn., Inc., Report No. 44: 1976.

Energy Management and Appliance Efficiency in Residences Nancy J. Smith, Interdisciplinary Group for Ecology, Development and Energy, Los Angeles, Calif.

Engineering and Design Analysis of an Energy Efficient Structure Utilizing Black Liquid Solar Collectors for Heating Dr. Tom Abeles and Martin Lunde, International Energy Conference, Tuscon, Arizona, Oct. 1977.

Ground Temperature Distribution with a Floor Panel Heating System A.B. Algren, ASHRAE Journal, pp. 111-119, May 1948.

The Effect of Ground Cover on Earth Temperature T. Kusuda, Alternatives in Energy Conservation: The Use of Earth Covered Buildings, NSF/RA — 760006, pp. 279-303, 1975.

index

Italic numbers refer to illustrations.

D

E

F

G

H

I

J

K

L

M

R

S

T

U